SpringerBriefs in Education

We are delighted to announce SpringerBriefs in Education, an innovative product type that combines elements of both journals and books. Briefs present concise summaries of cutting-edge research and practical applications in education. Featuring compact volumes of 50 to 125 pages, the SpringerBriefs in Education allow authors to present their ideas and readers to absorb them with a minimal time investment. Briefs are published as part of Springer's eBook Collection. In addition, Briefs are available for individual print and electronic purchase.

SpringerBriefs in Education cover a broad range of educational fields such as: Science Education, Higher Education, Educational Psychology, Assessment & Evaluation, Language Education, Mathematics Education, Educational Technology, Medical Education and Educational Policy.

SpringerBriefs typically offer an outlet for:

- An introduction to a (sub)field in education summarizing and giving an overview of theories, issues, core concepts and/or key literature in a particular field
- A timely report of state-of-the art analytical techniques and instruments in the field of educational research
- A presentation of core educational concepts
- An overview of a testing and evaluation method
- A snapshot of a hot or emerging topic or policy change
- An in-depth case study
- A literature review
- A report/review study of a survey
- An elaborated thesis

Both solicited and unsolicited manuscripts are considered for publication in the SpringerBriefs in Education series. Potential authors are warmly invited to complete and submit the Briefs Author Proposal form. All projects will be submitted to editorial review by editorial advisors.

SpringerBriefs are characterized by expedited production schedules with the aim for publication 8 to 12 weeks after acceptance and fast, global electronic dissemination through our online platform SpringerLink. The standard concise author contracts guarantee that:

- an individual ISBN is assigned to each manuscript
- each manuscript is copyrighted in the name of the author
- the author retains the right to post the pre-publication version on his/her website or that of his/her institution

More information about this series at https://link.springer.com/bookseries/8914

John T. E. Richardson

The Legibility of Serif and Sans Serif Typefaces

Reading from Paper and Reading from Screens

John T. E. Richardson
Institute of Educational Technology
The Open University
Milton Keynes, Buckinghamshire, UK

ISSN 2211-1921	ISSN 2211-193X (electronic)
SpringerBriefs in Education
ISBN 978-3-030-90983-3	ISBN 978-3-030-90984-0 (eBook)
https://doi.org/10.1007/978-3-030-90984-0

© The Author(s) 2022. This book is an open access publication.

Open Access This book is licensed under the terms of the Creative Commons Attribution 4.0 International License (http://creativecommons.org/licenses/by/4.0/), which permits use, sharing, adaptation, distribution and reproduction in any medium or format, as long as you give appropriate credit to the original author(s) and the source, provide a link to the Creative Commons license and indicate if changes were made.

The images or other third party material in this book are included in the book's Creative Commons license, unless indicated otherwise in a credit line to the material. If material is not included in the book's Creative Commons license and your intended use is not permitted by statutory regulation or exceeds the permitted use, you will need to obtain permission directly from the copyright holder.

The use of general descriptive names, registered names, trademarks, service marks, etc. in this publication does not imply, even in the absence of a specific statement, that such names are exempt from the relevant protective laws and regulations and therefore free for general use.

The publisher, the authors and the editors are safe to assume that the advice and information in this book are believed to be true and accurate at the date of publication. Neither the publisher nor the authors or the editors give a warranty, expressed or implied, with respect to the material contained herein or for any errors or omissions that may have been made. The publisher remains neutral with regard to jurisdictional claims in published maps and institutional affiliations.

This Springer imprint is published by the registered company Springer Nature Switzerland AG
The registered company address is: Gewerbestrasse 11, 6330 Cham, Switzerland

Dedicated to
James Hartley

Acknowledgements

I am most grateful to Charles Bigelow, Bethan Bolton, Francisco Cano-Garcia, James Clough, Martyn Cooper, Tim Coughlan, Mary Dyson, Robert Edmunds, Matt Elcock, James Hartley, Jason Humphries, Woojung Kim, Agnes Kukulska-Hulme, Christopher Lightfoot, Fulvia Mainardis, Marc Marschark, Rod McDonald, Kate Nation, Marta Novello, Bart Rienties, Norbert Schwarz, Ormond Simpson, Rebecca Treiman, Nicholas Wade, Sue Walker, Arnold Wilkins, Jiyeon Wood, and Gesualdo Zucco for their comments and advice. Any errors, inaccuracies or omissions are my own responsibility.

I am also very grateful to James Clough for his kind permission to make use of his photograph of the inscription in honour of Titus Annius Luscus in Fig. 1.3, to the Medical Research Council, as part of UK Research and Innovation, for its kind permission to make use of Fig. 2.1, and to the Philip's Division of Octopus Publishing Group and Sue Walker for their kind permission to make use of the photograph of pages from Nellie Dale's *Reader* in Fig. 7.1.

Milton Keynes, UK John T. E. Richardson
September 2021

Contents

1	**Introduction**	1
	1.1 The Origins of this Book	1
	1.2 Serif Typefaces	2
	1.3 Sans Serif Typefaces	4
	1.4 Review Methodology	8
	1.5 Conclusions	10
2	**Concepts and Research Methods**	11
	2.1 Concepts	11
	2.2 Objective Methods for Measuring the Legibility of Typefaces	12
	2.3 Subjective Methods for Measuring the Legibility of Typefaces	14
	2.4 The Size of Typefaces	15
	2.5 Conclusions	17

Part I Reading from Paper

3	**"Everybody Knows": Reading from Paper**	21
	3.1 Attitudes of Typographers	21
	3.2 Dissenting Voices	22
	3.3 Are Serifs Purely Decorative?	24
	3.4 Conclusions	25
4	**The Legibility of Letters and Words**	27
	4.1 Reading Letters and Words in Serif and Sans Serif Typefaces	27
	4.2 The "Stripiness" of Printed Words	29
	4.3 Confusions Among Letters in Serif and Sans Serif Typefaces	31
	4.4 Measuring Visual Acuity	32
	4.5 Conclusions	33

5	**Reading and Comprehending Text**		35
	5.1	Reading Text in Serif and Sans Serif Typefaces	35
	5.2	Comprehending Text in Serif and Sans Serif Typefaces	36
	5.3	The Connotative Meaning of Typefaces	38
	5.4	Connotations of Serif and Sans Serif Typefaces	39
	5.5	Conclusions	41
6	**Reading in Context**		43
	6.1	The Importance of Context	43
	6.2	Serif and Sans Serif Typefaces in Newspaper Headlines	44
	6.3	Wheildon's Research	46
	6.4	More Recent Research	48
	6.5	Conclusions	51
7	**Younger and Older Readers**		53
	7.1	Younger Readers	53
	7.2	Burt and Kerr's Research	54
	7.3	Zachrisson's Research	55
	7.4	Other Research with Children	56
	7.5	Letter Reversals	59
	7.6	Older Readers	60
	7.7	Conclusions	61
8	**Readers with Disabilities**		63
	8.1	Readers with Visual Impairment	63
	8.2	Shaw's Research	64
	8.3	Children in Special Education	66
	8.4	Readers with Congenital Visual Impairment	68
	8.5	Readers with Acquired Visual Impairment	68
	8.6	Readers with Aphasia	71
	8.7	Readers with Dyslexia	71
	8.8	Conclusions	73
9	**General Conclusions to Part I**		75
	9.1	Key Findings from Part I	75
	9.2	Preferences and Connotations	76
	9.3	Implications for Previous Assumptions	77
	9.4	The American Psychological Association's Current Position	77
	9.5	Conclusions	79

Part II Reading from Screens

10	**"Everybody Knows": Reading from Screens**		83
	10.1	Introduction	83
	10.2	Legibility of Serif and Sans Serif Typefaces Using Older Technology	85

	10.3 Issues with Screen Technology	88
	10.4 Conclusions	89
11	**The Legibility of Letters and Words**	**91**
	11.1 Reading Letters and Words in Serif and Sans Serif Typefaces	91
	11.2 The "Stripiness" of Words Displayed on Screens	93
	11.3 Confusions Among Letters in Serif and Sans Serif Typefaces	94
	11.4 Conclusions	95
12	**Reading and Comprehending Text**	**97**
	12.1 Reading Text in Serif and Sans Serif Typefaces	97
	12.2 Comprehending Text in Serif and Sans Serif Typefaces	99
	12.3 Rapid Serial Visual Presentation	102
	12.4 Reading Material on Handheld Devices and Smartphones	103
	12.5 Connotations of Serif and Sans Serif Typefaces	104
	12.6 Conclusions	106
13	**Readers with Disabilities**	**109**
	13.1 Readers with Visual Impairment	109
	13.2 Readers with Dyslexia	109
	13.3 Readers with Age-Related Macular Degeneration	110
	13.4 Conclusions	112
14	**Reading Text in Internet Browsers**	**113**
	14.1 The Legibility of Serif and Sans Serif Typefaces in Internet Browsers	113
	14.2 The Research of Bernard and Colleagues	114
	14.3 Subsequent Research	117
	14.4 Conclusions	121
15	**General Conclusions to Part II**	**123**
	15.1 Key Findings from Part II	123
	15.2 Preferences and Connotations	125
	15.3 Implications for Previous Assumptions	126
	15.4 Conclusions	127
16	**Coda: Lessons Learned**	**129**

References 133

Author Index 151

Subject Index 155

Typeface Index 159

List of Figures

Fig. 1.1	Examples of common serif typefaces (Baskerville, Garamond, Palatino, and Times New Roman) and common sans serif typefaces (Arial, Comic Sans, Tahoma, and Verdana)	3
Fig. 1.2	The surviving inscription on the base of Trajan's Column. Licensed under the Creative Commons Attribution–Share Alike 3.0 Unported (CC BY-SA 3.0), https://commons.wikimedia.org/wiki/File:Base_columna_trajana.jpg	3
Fig. 1.3	An inscription in honour of Titus Annius Luscus. Reproduced by kind permission of James Clough from https://articles.c-a-s-t.com/letter-hunting-in-italy-2-e7b51cd821a6	5
Fig. 1.4	A typical Snellen-type chart, used to evaluate visual acuity. Licenced under the Creative Commons attribution–Share Alike 3.0 Unported (CC BY-SA 3.0), https://commons.wikimedia.org/w/index.php?curid=4262200	7
Fig. 2.1	Key concepts in the measurement of typefaces. From *Effects of printing types and formats on the comprehension of scientific journals* (Applied Psychology Unit Report No. 346), by E. C. Poulton, 1959. UK Medical Research Council, Applied Psychology Unit. Used by kind permission of the Medical Research Council, as part of UK Research and Innovation	16
Fig. 7.1	Two pages from one of Dale's *Readers*. From *The Dale readers: Infant reader* (new ed.), by N. Dale. 1902. George Philip & Son. In the original, the illustrations and some letters in the sans serif headings were rendered in colour. Reproduced by kind permission of the Philip's Division of Octopus Publishing Group and Sue Walker from http://www.bookdata.kidstype.org/database/database/getImage?id=1018	54

Fig. 10.1　Examples of common serif typefaces (Times New Roman and Georgia) and common sans serif typefaces (Arial and Verdana) for displaying text on screens 84

Chapter 1
Introduction

1.1 The Origins of this Book

Many interesting research projects begin with an apparently simple question. The question with which this project began arose in the context of managing an online course.

In 2008, I and a colleague at the UK Open University were charged with designing and implementing a postgraduate distance-learning course with the title *Accessible Online Learning: Supporting Disabled Students*. The course was to be taken entirely online and would contribute to the University's master's programme in Online and Distance Education. All the course material was available online, either from a dedicated website or through the University library's online resources. In particular, the course textbook (Seale, 2006) was available free for students as an e-book. The course would run annually from September to January and would be equivalent to one quarter of a year's full-time study.

We recruited four experienced associate lecturers to serve as tutors for the course. They were each assigned 10–20 students, and their duties were to moderate online discussion forums among the group of students and to assess the assignments submitted by each student. The assignments were to be submitted in Microsoft Word format through a dedicated online system, and the students were given advice and instructions with regard to essay structure and referencing. The tutors provided an overall evaluation and mark (out of 100%) on each of the assignments using a separate form, and they provided more specific feedback in the margins of the assignment using Word's "comment" facility.

During the first presentation of the course in 2008–2009, one of the tutors suggested that students should be required to submit their assignments using a sans serif typeface, on the basis that "everybody knew" that sans serif typefaces were easier to read on screen and that this would make the task of evaluating the students' assignments more efficient. This seemed to overlook a number of points regarding the process of learning and assessment:

© The Author(s) 2022
J. T. E. Richardson, *The Legibility of Serif and Sans Serif Typefaces*,
SpringerBriefs in Education, https://doi.org/10.1007/978-3-030-90984-0_1

- Many students were using the default typeface in Microsoft Word to type their assignments. At the time, this was the serif typeface Times New Roman.
- The idea that the appearance of written work could be changed to suit the potential readers' abilities, skills, and preferences was introduced during the course.
- A tutor who downloaded an assignment to provide feedback could change the appearance of the assignment to suit their own abilities, skills, and preferences.
- Tutors might choose to evaluate assignments on screen, or they could instead choose to print off the assignments to read and evaluate them in hard copy.

The last point raised the question of the legibility of serif and sans serif typefaces when they were used to produce material intended to be read on paper or other hard surfaces. Here, a cursory view of the literature suggested that "everybody knew" that serif typefaces were easier to read on paper than were sans serif typefaces. Of course, the fact that everybody knows something is no guarantee that it is true. Until the time of the Pythagorean School in the sixth century BC, "everybody knew" that the earth was flat; and until the time of Copernicus in the sixteenth century AD, "everybody knew" that the sun rotated around the earth. Nowadays, we expect such matters to be determined by empirical evidence, not by majority opinion. This book is concerned with the empirical evidence concerning the relative legibility of serif typefaces and sans serif typefaces: Part I is concerned with their legibility in "hard copy" (i.e., when presented on paper or other hard surfaces), and Part II is concerned with their legibility when presented on computer monitors or other screens.

1.2 Serif Typefaces

There are many dimensions on which typefaces can vary, but this book is concerned with the legibility of typefaces with and without serifs. A serif is "a short, light line projecting from the main stroke of a letter" (*Chicago Manual of Style*, 2003, p. 837) that often takes the form of a small finishing stroke at the top or bottom of a letter. (In English-speaking countries, typographers have variously spelled the word *ceref*, *ceriph*, *seriff*, *seriph*, *seryph*, *surriph*, *surripse*, *surryph*, or *syrif*: Mosley, 1999, pp. 53, 55.) One feature of typefaces with serifs is that the main strokes constituting each letter are often of varying thickness. The left-hand panel of Fig. 1.1 shows some examples of common serif typefaces (Baskerville, Garamond, Palatino, and Times New Roman) as they are currently rendered in Microsoft Word. (The example sentence is a pangram—a single sentence that uses all 26 letters in the English alphabet—which is often used as a typing exercise.)

Serifs have been noted in Greek inscriptions from the fourth century BC, but they became widely adopted during the Roman Empire (from 27 BC to AD 476) (Mosley, 1999, p. 18), when they are generally thought to have resulted from the practices of Roman masons (Bringhurst, 2019, pp. 119–120). It is likely that the latter used a flat or square-edged brush to draft symbols on pieces of stone before carving them with chisels, and the serifs were left at the ends of the brushstrokes

1.2 Serif Typefaces

Baskerville: The quick brown fox jumps over the lazy dog.	Arial: The quick brown fox jumps over the lazy dog.
Garamond: The quick brown fox jumps over the lazy dog.	Comic Sans: The quick brown fox jumps over the lazy dog.
Palatino: The quick brown fox jumps over the lazy dog.	Tahoma: The quick brown fox jumps over the lazy dog.
Times New Roman: The quick brown fox jumps over the lazy dog.	Verdana: The quick brown fox jumps over the lazy dog.

Fig. 1.1 Examples of common serif typefaces (Baskerville, Garamond, Palatino, and Times New Roman) and common sans serif typefaces (Arial, Comic Sans, Tahoma, and Verdana)

(Catich, 1991). By the beginning of the Common Era, Roman inscriptions used a standard alphabet of capital letters in which serifs were a characteristic feature, and many examples have survived in their original locations or else in museums to the present day. A commonly cited example is the inscription on the base of the column that was completed in AD 113 to commemorate the emperor Trajan's victory in the Dacian Wars, shown in Fig. 1.2. The column itself survives largely intact in the otherwise ruined remains of Trajan's forum near the Piazza Venezia in the centre of modern Rome.

Smaller versions of these letters were used when writing books. However, towards the end of the eighth century AD, a simplified version of this alphabet, nowadays known as the "Carolingian minuscule," was introduced across the Holy Roman Empire as a standard handwritten script for rendering the Vulgate Bible in Latin. This incorporated strokes that extended above or below the main body of each letter but

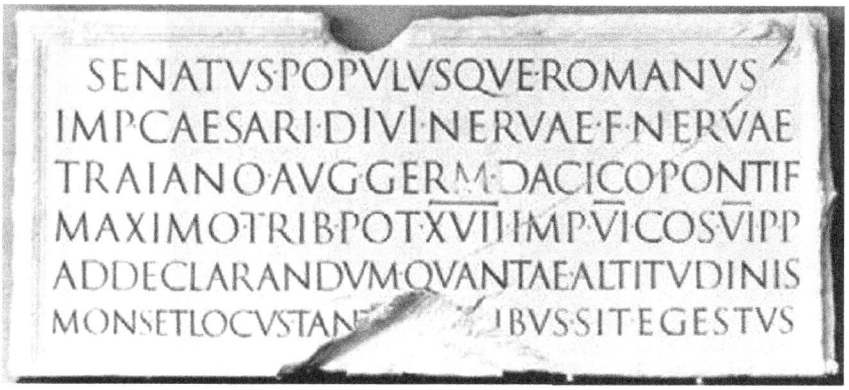

Fig. 1.2 The surviving inscription on the base of Trajan's Column. Licensed under the Creative Commons Attribution–Share Alike 3.0 Unported (CC BY-SA 3.0), https://commons.wikimedia.org/wiki/File:Base_columna_trajana.jpg

retained the use of serifs. In the early fifteenth century, Italian calligraphers combined the Roman capitals with the Carolingian minuscule; these were used as the basis for the earliest Western typefaces and evolved into the combination of uppercase and lowercase alphabets that is used in Western countries today (see Bigelow, 1981, for a more detailed account, of which this is mainly a summary). Due to the origins of their capital letters in inscriptions from the Roman Empire, serif typefaces are sometimes referred to as *Roman* or *roman*. For instance, Times New Roman was developed by Stanley Morison, working for the Monotype foundry, for use in the London newspaper, *The Times*, in 1932. A rival printing company, Linotype, developed a similar typeface known as "Times Roman" (or even just as "Times").

Several theories have been put forward regarding why serifs should have survived in modern typography:

- Early researchers sometimes claimed that serifs provided additional visual cues to enable readers to direct their gaze at successive words in a line of text. This idea can be found even in some modern accounts. However, the work of Hering (1879) and Lamare (1892) showed that the eye movements of experienced readers consist not in a continuous horizontal gaze but in a series of discrete fixations separated by jumps or "saccades". (This finding is often erroneously attributed to their colleague, Louis Émile Javal: see Wade & Tatler, 2008, for discussion of this issue.)
- Other early researchers argued that serifs helped to overcome the harmful effects of "irradiation" (see Pyke, 1926, pp. 21, 99–101, for examples). The latter is a well-established optical illusion whereby a dark figure that is presented against a light visual field appears to be larger than an otherwise identical light figure that is presented against a dark visual field. Taylor (1934) claimed that irradiation explained why letters printed in serif typefaces were harder to read when shown in white print against a black background than when shown in black print against a white background. Nevertheless, the relevance of this phenomenon to the legibility of typefaces is otherwise unclear.
- Robinson et al. (1971) hypothesised that serifs facilitated the operation of line detectors in the human visual system. They found evidence in support of this idea using a computer model of visual processing. Nevertheless, computational models which assume that specific and unique brain cells are dedicated to the detection of lines or other features have since been criticised in favour of connectionist models which assume that groups of brain cells function as a distributed network (e.g., Schiffman, 2000, pp. 83–85, 163–166).

1.3 Sans Serif Typefaces

Sans serif typefaces are presented without serifs. (In English-speaking countries, typographers have variously used the expressions *sans-ceriph*, *sans-serif*, *sans-surryph*, *sanserif*, and *sansserif*: Mosley, 1999, p. 53.) In contrast to serif typefaces, the strokes constituting each letter are often of constant thickness. The right-hand

1.3 Sans Serif Typefaces

panel of Fig. 1.1 shows examples of common sans serif typefaces (Arial, Comic Sans, Tahoma, and Verdana) as they are currently rendered in Microsoft Word.

Early Greek and Etruscan inscriptions routinely employed a sans serif style (Mosley, 1999, p. 17), and they were widely adopted during the Roman Republic (from 509 to 27 BC) (Bringhurst, 2019, p. 261). In contrast to serif inscriptions, relatively few examples survive today (Lightfoot, 2009), partly because of the poorer quality of the material in which they were inscribed (wood or local stone, rather than marble) and partly because from time to time the Republican authorities discouraged the construction of inscribed monuments. Nevertheless, the National Archeological Museum in Aquileia in north-eastern Italy contains many Republican inscriptions using sans serif capital letters (Clough, 2015), and about 150 of these are displayed in the online Lupa database (http://lupa.at). Figure 1.3 shows an inscription dating from the middle of the second century BC that was discovered in the area of the Roman forum in Aquileia in 1995. It came from a monument (now lost) to Titus Annius Luscus, who was elected as praetor in 156 and consul in 153 BC. Clough

Fig. 1.3 An inscription in honour of Titus Annius Luscus. Reproduced by kind permission of James Clough from https://articles.c-a-s-t.com/letter-hunting-in-italy-2-e7b51cd821a6

(2020) provided a more detailed discussion of this inscription. (I am grateful to James Clough for permission to reproduce his photograph of the inscription here.)

Sans serif inscriptions had a brief revival in the fifteenth century, when they were used to decorate buildings and monuments in a number of Italian cities (Clough, 2020; Gray, 1960). The origins of the style are a matter of debate (Stiff, 2005). Nevertheless, these developments had little or no effect on the evolution of early typefaces. Instead, early typographers used serif typefaces modelled on surviving inscriptions from Imperial Rome, and these had become widely adopted by the end of the eighteenth century.

During the latter part of the eighteenth century and the early nineteenth century, sans serif inscriptions became popular on monuments, public buildings, and garden features in the United Kingdom, where they were seen as being more natural or primitive than serif inscriptions. Between 1800 and 1820, hand-etched sans serif letters were used in publicity materials printed for commercial signwriters or engravers and on the title pages of British catalogues of antiquities, while lowercase sans serif letters were engraved from around 1810. At this time, the style was described as "Egyptian", and it was this term that was used to describe the first uppercase sans serif typeface produced in 1816. However, the first sans serif typeface produced in both uppercase and lowercase in 1832 was described as "sans-serif", and variations of this term were used thereafter. Such typefaces were occasionally referred to as "antique" or "grotesque", and the equivalent terms were adopted by typographers in France and Germany, respectively, later in the nineteenth century (see Mosley, 1999, for a more detailed account, of which this paragraph is mainly a summary; see also Mosley, 2007).

Possibly in reaction to the popularity of sans serif styles, some printers devised a variant of the serif style in which the serifs were of a similar thickness to the letters' main strokes. These were used in wood blocks from 1810 and in metal type from 1817. Somewhat confusingly, these typefaces were referred to as "antique" in the United Kingdom and the United States. However, they were also occasionally referred to as "Egyptian", and the equivalent terms were adopted by typographers in France and Germany. In English-speaking countries, they are nowadays referred to as "slab serif" typefaces (Mosley, 1999, pp. 42, 56; 2007). Figure 1.4 shows a typical example of a Snellen chart, used to assess visual acuity (to be discussed in Sect. 4.3). This uses slab serif symbols, each constructed within a 5×5 grid.

From the 1830s, sans serif typefaces were widely used in commercial printing (Mosley, 1999, p. 43). Initially, they were mainly used for display purposes (Kinross, 1992, pp. 28–29; McLean, 1980, p. 64). Nowadays, as Perea (2013) noted, sans serif typefaces are widely used in many countries for public direction signs, although it caused some controversy when they were introduced for a new motorway (freeway) system in the United Kingdom in 1959 (see Lund, 1999, pp. 126–147). Some foundries in the United Kingdom adopted the term "gothic" for sans serif typefaces around the middle of the nineteenth century, and this term became generally used in the United States (Mosley, 1999, pp. 55–56). There was much global interchange in the evolution of typography, and the distinction between serif and sans serif typefaces appears to be universal in countries that have adopted a Western alphabet (Ovink,

1.3 Sans Serif Typefaces

Fig. 1.4 A typical Snellen-type chart, used to evaluate visual acuity. Licenced under the Creative Commons attribution–Share Alike 3.0 Unported (CC BY-SA 3.0), https://commons.wikimedia.org/w/index.php?curid=4262200

1938, pp. 188–228). More recently, this has been compounded by the hegemony of word-processing software originating in the United States.

The distinction between serif and sans serif typefaces applies to most typefaces that are designed for reading over long stretches of text. It is less applicable to other kinds of typeface, such as *display* typefaces that are designed to attract attention (for instance, in advertisements or logos) and *cursive* typefaces that are intended to mimic handwriting.

1.4 Review Methodology

This book reports the findings of a systematic review of research comparing the legibility of serif typefaces and sans serif typefaces. As Uman (2011, p. 57) explained, "narrative reviews" (in other words, conventional literature reviews)

> can often involve an element of selection bias. They can also be confusing at times, particularly if similar studies have diverging results and conclusions. Systematic reviews, as the name implies, typically involve a detailed and comprehensive plan and search strategy derived a priori, with the goal of reducing bias by identifying, appraising, and synthesizing all relevant studies on a particular topic.

Readers may be aware of the systematic reviews from the Cochrane Collaboration, an international organisation created in 1993 to focus on health-related issues. A complementary organisation, the Campbell Collaboration, was established in 2000 to focus on social-related issues (see Noonan & Bjørndal, 2010).

The first stage in any systematic review is to identify a search strategy based upon key terms. This can be simple or complex and in clinical research can involve the specification of inclusion or exclusion criteria. In this case, the aim was to identify all previous studies which had endeavoured to compare the legibility of serif and sans serif typefaces. It was therefore decided to use the single key term *serif*. "Legibility" is intrinsically a psychological concept with educational applications, but it was recognised that the legibility of serif and sans serif typefaces might vary across particular clinical populations. Accordingly, the following online databases were deemed relevant:

- APA PsycInfo (https://www.apa.org/pubs/databases/psycinfo; formerly PsycINFO) contains approximately 5 million records, mainly relating to peer-reviewed publications. It subsumes the journal *Psychological Abstracts*, which went back to 1894, but it also contains some earlier publications.
- ERIC (https://eric.ed.gov/) is the bibliographic database of the US Education Resources Information Center. It contains more than 1.6 million records of education-related materials. The collection was initiated in 1966, although it contains some earlier material. In the past, authors could use it as a repository for their own material, and so a proportion of the records relates to "grey" literature that has not been peer-reviewed. However, with effect from January 2016, ERIC introduced a selection policy that limited new records to material that had undergone some kind of review process.
- MEDLINE (https://www.nlm.nih.gov/medline/index.html) is the bibliographic database of the US National Library of Medicine. It contains more than 27 million records relating to journal articles in the life sciences with a focus on biomedicine.

All three databases are accessible through EBSCO Information Services, which means that they can be searched simultaneously to avoid duplicate results.

These three databases were searched repeatedly during the period 2019–2021 to find publications containing the term *serif* in their titles, abstracts, keywords, or metadata. This led to a high number of false positives mainly because "Serif" is a common first name and family name in Turkey. It was also suspected to lead to a high number of misses, because informally it was noted that some relevant sources were not covered by any of the three databases. The results were therefore used as a basis for the conventional procedures of backward searching and forward searching. The former refers to the examination of *previously* published sources cited by the obtained hits, while the latter refers to the examination of *subsequently* published sources that cite one or more of the obtained hits. This process was facilitated by employing the database Web of Science (https://clarivate.com/products/web-of-science/, formerly Web of Knowledge), which enables searching among 79 million cited or citing sources in the form of books, journals, and conference proceedings.

When there is sufficient commonality with regard to research methods, it is possible to integrate the quantitative findings of a systematic review using statistical techniques, thus yielding a single overall estimate of a difference, variation, or effect size. Such an approach is known as "meta-analysis". A classic example is the analysis of differences between men and women in their performance on particular cognitive tests or similar tasks (for example, see Caplan et al., 1997). Nevertheless, despite the apparently simple nature of the research question, the present literature review yielded extremely diverse methods of data collection and analysis, and this ruled out any formal statistical meta-analysis to integrate the research findings from the wide variety of studies that will be described.

The alternative approach is to rely on "vote counting" (sometimes known as the "box score" approach). For present purposes, different studies are sorted according to whether their results are statistically significant favouring serif typefaces, statistically significant favouring sans serif typefaces, or not statistically significant, and I shall invite readers to plump for the majority outcome across these three categories in particular tasks, in particular contexts, and in particular subject populations; in other words, I shall focus readers' attention on the most common finding or, as statisticians would say, the *modal* finding. Such an approach is not without its hazards (see Caplan, 1979; Maccoby & Jacklin, 1974, pp. 355–356), and it can in theory lead to misleading conclusions (see Hedges & Olkin, 1985, pp. 48–52). Nevertheless, it has the major advantage over the use of narrative review that the criteria and standards used for the selection and interpretation of individual studies have been made totally explicit, and hence the findings can be readily replicated. In fact, in most cases the results are sufficiently unambiguous that readers should have little difficulty sharing my conclusions.

The final methodological point is that there is no reason to think that typographical features have the same consequences when people are reading from paper or other hard surfaces and when they are reading from computer monitors or other screens. It follows that reviews which indiscriminately combine research on reading from paper with research on reading from screens (e.g., Chung, 2020) are unlikely to be informative. Accordingly, Part I of this book reviews the research literature regarding the legibility of serif typefaces and sans serif typefaces when they are used to generate material that is printed on paper, and Part II reviews the research literature regarding the legibility of serif typefaces and sans serif typefaces when they are used to produce material that is to be viewed on display screens or by means of other kinds of technology.

1.5 Conclusions

This chapter has introduced the distinction between serif and sans serif typefaces. There seem to be widespread assumptions about their relative legibility both on paper and on screens. The chapter also described the methodology of systematic review employed to address this issue.

Open Access This chapter is licensed under the terms of the Creative Commons Attribution 4.0 International License (http://creativecommons.org/licenses/by/4.0/), which permits use, sharing, adaptation, distribution and reproduction in any medium or format, as long as you give appropriate credit to the original author(s) and the source, provide a link to the Creative Commons license and indicate if changes were made.

The images or other third party material in this chapter are included in the chapter's Creative Commons license, unless indicated otherwise in a credit line to the material. If material is not included in the chapter's Creative Commons license and your intended use is not permitted by statutory regulation or exceeds the permitted use, you will need to obtain permission directly from the copyright holder.

Chapter 2
Concepts and Research Methods

2.1 Concepts

Since the introduction of movable type in Western countries during the fifteenth century, many thousands of different typefaces have been designed for use in printed material. A *typeface* can be expressed in several different *fonts* (bold, italic, etc.) by varying the weight, width, and style of individual characters. Since the seventeenth century, there has been an alternative use of *font* (and its variant *fount*) as a synonym for *typeface*, and this has become more common since the introduction of digital typography (Oxford University Press, n.d.). Nevertheless, for consistency, the words *typeface* and *font* will be used in their original senses in this book; thus, a *typeface* is comprised of a family of related *fonts*. In Sect. 1.2, for example, Fig. 1.1 showed eight different *typefaces*, and both the name of each typeface and the example sentence (the pangram) were shown in the typeface's regular *font*.

Typefaces are designed to be read, and thus an obvious research question is whether different typefaces vary in how legible they are for readers. *Readable* can be used as a synonym of *legible*, although there are technical definitions of both *legibility* and *readability* that go beyond their daily use. Some researchers have used "legibility" to refer to the recognition and the identification of individual letters or words and "readability" to refer to the reading and the understanding of connected prose. Others have devised "readability formulas" to measure the level of mental difficulty involved in reading specific material. Yet others have used "readability" to refer to the extent to which a typeface is subjectively appealing or comfortable to the reader. Even so, as Chomsky (1970) pointed out, in many current versions of English, *readable* is much more sharply restricted in meaning than "able to be read": it is instead often used to mean how easy, enjoyable, or engaging a work is to read, as in "This is a most readable novel". (Chomsky explained that this phenomenon was problematic for theories of transformational grammar.) Consequently, *legible* and *legibility* will be used throughout this book.

The legibility of typefaces is pertinent to a wide variety of everyday settings, but it is particularly relevant for the field of education. First, much of the information that is acquired by students is delivered in books, articles, or other printed documents presented either on paper or computer screens or in printed displays projected using PowerPoint or other software. Second, students often submit their work to be evaluated by their teachers or other assessors in the form of word-processed documents, which raises the issue of their legibility for those teachers and assessors, and which led to the question that gave rise to this project.

2.2 Objective Methods for Measuring the Legibility of Typefaces

Attempts to measure the legibility of printed material go back at least to the 1880s. Tinker (1963, pp. 5–7, 9–31) provided a useful summary of the relevant methods of investigation (see also Pyke, 1926, pp. 25–34; Reynolds, 1979; Zachrisson, 1965, pp. 44–69). The following list of methods is a paraphrase based mainly upon Tinker's account, but it covers most of the techniques that have been used to measure the legibility of printed material. Most of them can be applied equally to measure the legibility of material presented on computer monitors or other screens.

- **Short-exposure method**. Printed symbols are briefly presented (e.g., by means of a tachistoscope, which carefully controls the duration of a presentation using shutters and mirrors) to measure the speed or accuracy with which they can be perceived and reported.
- **Distance method**. Printed symbols are presented in clear view but at a distance from the observer. The material is then moved towards the observer in gradual steps to measure the furthest distance at which they can be perceived and reported correctly. A variant is where the observer gradually approaches the stimulus. Similar techniques to compare the legibility of different typefaces have been employed since the eighteenth century (Kinross, 1992, pp. 23–24).
- **Perceptibility in peripheral vision**. Printed symbols are presented to one side or the other of a central fixation point to measure the furthest horizontal distance at which they can be perceived and reported correctly. Similar effects can be obtained using the "focal variator" (Weiss, 1917), which uses a system of lenses to project a visual stimulus onto a ground glass screen to varying degrees out of focus.
- **Visibility threshold**. Printed symbols are viewed through two photographic filters with precise circular gradients of density which are rotated until the material can be perceived and reported correctly. The filters reduce the apparent brightness of the material and also lower the contrast between the material and its background (Luckiesh & Moss, 1935, 1942, pp. 71–79).
- **Reflex blink method**. The observer reads text, and the experimenter counts the number of involuntary eye blinks made during a standard observation interval. This assumes that the blink rate is reduced and the reader's progress faster with more legible text.

- **Rate of work**. This covers a variety of tasks including speed of reading, amount read in a specified time limit, the time taken to look up specific information in printed sources such as telephone numbers or functions in mathematical tables, and output in tasks involving visual discrimination.
- **Eye movements**. The observer is asked to read continuous text, and the experimenter measures the number of fixations and the number of jumps or saccades between successive fixations. This assumes that more legible text results in shorter fixations and fewer saccades.
- **Fatigue in reading**. This approach is concerned not with visual fatigue in reading per se, which has proved consistently difficult to measure; rather, legibility is defined in terms of the ease, accuracy, or efficiency of perceiving printed symbols while reading for understanding.

Of course, new technologies to measure legibility have been introduced over the years. For instance, to employ the short-exposure method, Cattell (1885) used a "gravity chronometer" in which printed material was obscured by a vertically sliding panel held in place by an electromagnet. On its release, the panel fell, and the material was visible for a brief period of time through a small window in the panel. Cattell found that both uppercase letters and lowercase letters varied considerably in their legibility. More sophisticated tachistoscopes became available in the early twentieth century. Since the 1970s, technology has included cathode-ray tubes and liquid crystal displays, and these will be discussed in Part II. Again, studies of eye movements in reading have become more popular with the use of computer-based eye-tracking devices, and these will also be discussed in Part II.

Not only have researchers adopted different methods for measuring the legibility of printed material, but they have presented their participants with different kinds of material: individual letters or other characters; letter strings that do not constitute words; individual words; sequences of unrelated words; strings of words that constitute grammatical sentences; or coherent grammatical prose. The materials towards the beginning of this list afford more opportunity to control the participants' behaviour, whereas the materials towards the end of the list are more akin to those encountered in everyday reading situations. As in other kinds of educational and psychological research, there is a trade-off between experimental rigour and "ecological validity" (i.e., whether the findings can be generalised to real-life settings).

In particular, Kinross (1992, p. 32) noted that most of the research carried out before the end of the nineteenth century had tested the recognisability of individual letters rather than the legibility of words or passages of text. As he pointed out, it took a change in the theoretical climate around 1900 for legibility to be interpreted as the comprehension of *meaning*: "not recognition, but reading." Tinker (1963) went so far as to propose the following conclusion: "Research dealing with individual letters or letters grouped in nonsense arrangement offers little that is important concerning the legibility of type faces. Satisfactory results are obtained by measuring speed of reading continuous, meaningful material" (pp. 65–66).

It is important to distinguish between the legibility of different typefaces and readers' familiarity with these typefaces. This is reflected in a study of binocular rivalry by Zachrisson (1965, pp. 128–131). The latter phenomenon occurs when a stimulus is presented to one eye but another stimulus is presented to the other eye: instead of seeing the stimuli superimposed on each other, most observers report seeing images of the stimuli alternating with each other. Zachrisson presented 28 students with individual words. In each case, the word was presented in the serif typeface Imprint to one eye and in the sans serif typeface semi-bold Grotesk to the other eye. The participants pressed one of two buttons to report which typeface they were seeing over a period of 3 min. The results showed a strong dominance of the serif typeface over the sans serif typeface, regardless of the eyes to which they were presented. Zachrisson repeated his study with 9-year-old children and found that the dominance of the serif typeface was much weaker. He took this to reflect the children's reduced familiarity with these letter forms. The implication is that the stronger ocular preference seen in adults was mainly due to their more extensive experience of reading documents (such as newspapers, magazines, reports, and books) in serif typefaces rather than to any inherent superiority in their legibility.

2.3 Subjective Methods for Measuring the Legibility of Typefaces

Researchers have also collected subjective reports from their participants concerning the legibility of different typefaces. Examples of different typefaces can be presented either individually or in groups of two or more for comparison. The self-reports can be collected either informally (for example, through interviews) or more formally (for example, through the use of rankings or rating scales). Pyke (1926, pp. 58–59) asked 60 participants to rank order eight typefaces (including one sans serif typeface, Lining Grotesque) in terms of their "relative merits". He found that the participants gave various reasons for their choices, and many found it difficult to differentiate between the typefaces on this basis. He found that there was a correlation of just +0.46 with the rank order of performance in a speed-of-reading test, and he concluded that the relationship between the two measures was unclear.

Tinker and Paterson (1942) asked a group of participants to arrange samples of ten different typefaces "in order from most legible to least legible" (p. 38). Tinker (1944) then obtained results on the legibility of the ten typefaces using the visibility threshold method, which he referred to as their "visibility". He had existing data on the legibility of the same typefaces using the distance method (which he referred to as their "perceptibility") as well as data on their legibility using their speed of reading. He found a high positive correlation between the ranks of their visibility and their perceptibility, which suggested that the two measurements had much in common. Nevertheless, their ranked visibility and perceptibility both showed a modest *negative* relationship with their ranked speed of reading. In addition, their judged legibility showed a high positive correlation with their ranked visibility and perceptibility.

However, consistent with the results obtained by Pyke (1926), their judged legibility showed only a correlation of +0.33 with their ranked speed of reading.

These findings could be interpreted in a number of different ways. Tinker himself argued that both visibility and perceptibility at a distance represented abnormal and artificial reading situations, even though they were highly correlated with judged legibility. Instead, he argued that speed of reading constituted the best possibility as a measure of legibility since it provided "measurement in a normal, ordinary reading situation" (Tinker, 1944, pp. 393–394). In addition, he claimed that subjective judgments of legibility "can only be considered as an expression of preference which may be employed to advantage in a practical way for the guidance of printers when there is a choice to be made between equally readable typographical arrangements" (pp. 394–395). Even so, Tinker's results imply that the techniques listed in Sect. 2.2 do not simply constitute alternative measures of a single construct of legibility. Any research findings with regard to the legibility of serif and sans serif typefaces therefore need to be qualified by a clear explanation of the measure or measures of legibility on which they are based, and this practice will be adopted in this book.

Participants' preferences might not be a reliable indicator of the objective legibility of different typefaces, but they may well have practical consequences. Song and Schwarz (2008b) carried out three studies in which the participants read instructions for carrying out a particular task printed either in a plain sans serif typeface (Arial in all three studies) or in an elaborate cursive typeface (Brush455 BT or Mistral in different studies). In all three experiments, the sans serif typeface was rated as easier to read than the cursive typeface, but there was no difference in the participants' memory for particular details in the instructions. The participants who read the instructions in the cursive typeface reported that the task would take more time, would feel less fluent and natural, and would require more skill, and that they were less willing to engage in the task than were the participants who read the instructions in the sans serif typeface. Song and Schwarz concluded that the participants had mistaken the ease of processing the instructions as indicating the ease with which the relevant tasks could themselves be executed. Song and Schwarz (2008a) showed that the same manipulation affected how participants answered distorted and undistorted questions based on their general knowledge.

2.4 The Size of Typefaces

It might seem plausible that the legibility of different typefaces depends on their size. In fact, Legge and Bigelow (2011) showed that legibility was essentially constant across the range of type sizes that readers might encounter in books, magazines, and newspapers. Nevertheless, comparing the physical size of different typefaces is not a straightforward matter.

Traditionally, the overall height of typefaces (technically known as their *body size*) has been expressed in terms of points, where one point is approximately equal to 0.35 mm. However, the size of typefaces is also expressed in terms of the dimensions

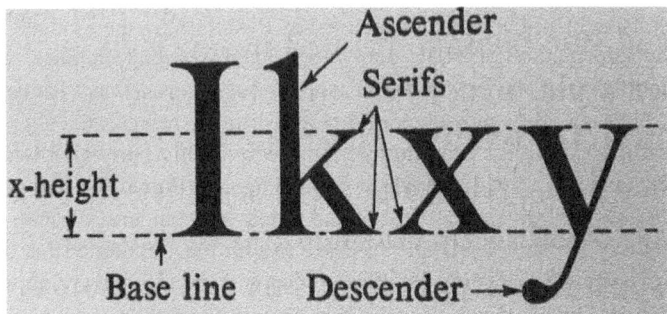

Fig. 2.1 Key concepts in the measurement of typefaces. From *Effects of printing types and formats on the comprehension of scientific journals* (Applied Psychology Unit Report No. 346), by E. C. Poulton, 1959. UK Medical Research Council, Applied Psychology Unit. Used by kind permission of the Medical Research Council, as part of UK Research and Innovation

of lowercase letters. Some lowercase letters have features that extend above their main parts (e.g., *b* and *d*); these are called *ascenders*. Others have features that extend below their main parts (e.g., *p* and *q*); these are called *descenders*. The *x-height* of a typeface is the height of lowercase letters that do not have either ascenders or descenders (such as the letter *x* itself). Finally, the *cap-height* of a typeface is the height of capital (or uppercase) letters, which may or may not be the same as the height of ascenders. Key concepts in the measurement of typefaces are summarised in Fig. 2.1.

The body size of a typeface is thus made up of its x-height and the combined heights of the ascenders and descenders, plus small margins above the tops of ascenders and below the bottoms of descenders. (The latter are known as *leading*, pronounced "ledding". This term originated in the practice of using thin strips of lead to separate lines of text in order to increase the vertical space between them. Such terminology was developed in the age of movable type, but it has been carried over into electronic printing, where it is also known as *interline spacing*.) When comparing different styles of typeface, researchers have often matched them on the basis of their body size. Nevertheless, Poulton (1972) noted that this does not equate the sizes of the individual letters, as measured by their x-height. In general, it is often not possible to match pairs of typefaces simultaneously on the basis of *both* their body size *and* their x-height.

Poulton (1972) simulated the situation of a shopper looking for a particular item in the list of ingredients on a package of food. Because food containers are often quite small, the typefaces used for lists of ingredients are themselves usually quite small. Poulton tried to determine the minimum legible size of lowercase letters printed in one of two serif typefaces (Times New Roman and Perpetua) or in one sans serif typeface (Univers). He asked a total of 264 adult volunteers to find a designated target word within each of 15 lists of food ingredients, for which they were allowed 25 s. The number of target words found within this time limit was used as a measure of legibility.

Poulton found that performance markedly declined when the x-height of a typeface was less than 1.2 mm. He also found that Times New Roman and Perpetua yielded similar results, even though the latter's body size was more than 30% greater than that of the former. He inferred that body size was not an important determinant of legibility. Performance was significantly poorer with Univers than with either Times New Roman or Perpetua, but not when the x-height of the two latter typefaces was reduced photographically to match that of the former. Poulton concluded that Univers was less legible than the other typefaces because of its smaller x-height and not because of the absence of serifs. In fact, typographers have believed for a long time that the visual impact of lowercase letters is determined by their x-height rather than by their body size (Craig, 1971, p. 24; Williamson, 1966, p. 37).

2.5 Conclusions

This chapter clarified the distinction between *typefaces* and *fonts* and that between *legibility* and *readability*. It described the variety of objective methods that have been used to measure the legibility of printed material and the different ways of collecting subjective reports from participants regarding the legibility and other properties of presented material. Most of these techniques have been taken over into research on reading from screens. Finally, this chapter described how typographers define the size of typefaces and discussed which aspects of the size of typefaces are likely to affect the legibility of material.

Open Access This chapter is licensed under the terms of the Creative Commons Attribution 4.0 International License (http://creativecommons.org/licenses/by/4.0/), which permits use, sharing, adaptation, distribution and reproduction in any medium or format, as long as you give appropriate credit to the original author(s) and the source, provide a link to the Creative Commons license and indicate if changes were made.

The images or other third party material in this chapter are included in the chapter's Creative Commons license, unless indicated otherwise in a credit line to the material. If material is not included in the chapter's Creative Commons license and your intended use is not permitted by statutory regulation or exceeds the permitted use, you will need to obtain permission directly from the copyright holder.

Part I
Reading from Paper

Chapter 3
"Everybody Knows": Reading from Paper

3.1 Attitudes of Typographers

Section 1.1 noted the uncritical acceptance of the view that serif typefaces were easier to read on paper and other hard surfaces than were sans serif typefaces. This view has traditionally been adopted by typographers and typography educators (see, e.g., Craig, 1971, pp. 123–125; Hamai, 1986, p. 7; McLean, 1980, p. 44; Williamson, 1966, p. 109). Such authors have often relied on their attitudes and experience (and, sometimes, their authority and influence), but they show little awareness that the issue might be subjected to formal empirical research. The most extreme position was adopted by Morison (1959), who had developed the serif typeface Times New Roman in 1932 for the London newspaper, *The Times* (see Sect. 1.2); Morison insisted that "the serif is essential to the reading of alphabetical composition in long passages and consecutive pages" (p. xi), but he provided no evidence for this assertion. In short, as far as 20th-century typography was concerned, it was definitely a matter of "everybody knew" that serif typefaces were easier to read than sans serif typefaces when printed on paper.

These views have often been uncritically incorporated into the guidelines for potential authors that have been produced by journal editors and publishers. Such guidelines tend to include assumptions about the legibility of serif and sans serif typefaces, typically without providing any arguments or evidence in support of those assumptions. One characteristic example can be found in the sixth (2010) edition of the *Publication Manual of the American Psychological Association*. In a section headed "Preparing the Manuscript for Submission," readers were instructed as follows:

> A *serif* typeface ... is preferred for text because it improves readability and reduces eye fatigue. (A *sans serif* type may be used in figures, however, to provide a clean and simple line that enhances the visual presentation.) (American Psychological Association, 2010, pp. 228–229).

This is contrary to good practice in psychology and the social sciences, where it is nowadays expected that assertions of this kind will be supported by citations of published (and, ideally, peer-reviewed) research, not based on presumed authority or ex cathedra pronouncements. Even so, the Association's guidelines were adopted by the American Educational Research Association and by many other organisations both in the United States and around the world.

Another example can be found in *Merriam-Webster's Manual for Writers and Editors* (1998), although this did at least allude to the existence of empirical evidence on the issue:

> Serif faces are somewhat easier to read in blocks or paragraphs of text than sans-serif faces. . . . Studies of typeface legibility have tended to demonstrate that standard serif typefaces can be read somewhat more easily and quickly than standard sans-serif typefaces. (pp. 329–330).

However, since no such studies were explicitly cited, it is impossible for an interested but sceptical reader to determine whether or not the *Manual*'s account was accurate.

3.2 Dissenting Voices

During the twentieth century, there were few dissenting voices from this dominant view within typography. Although he had been one of Morison's colleagues, Dreyfus (1985, p. 19) stated: "The outcome of many experiments indicates there is no statistically significant difference between the legibility of a wide variety of text types, even between seriffed and unseriffed types." Unfortunately, he failed to specify which "experiments" he had in mind.

In the twenty-first century, dissenting voices have tended to come from the editors of journals in medicine and bioscience. They would, of course, be comfortable with the idea that the legibility of different typefaces could be the subject of empirical research, but their comments suggest that they typically lacked specialised knowledge of this research literature. In 2004, for example, the *Journal of Psychopharmacology* moved from a serif typeface to a sans serif typeface, which the editor claimed (once again without citing any sources) "should improve visual impact and reading ease of the Journal" (Nutt, 2004, p. 5).

In 2011, the editor of the Spanish journal *Revista Española de Anestesiología y Reanimación*, notified its readers of a "new look" for the journal (Errando Oyonarte, 2011a). (It specialises mainly in anaesthesiology, resuscitation, and pain management.) He mentioned a number of changes in the appearance and the format of the journal with the aim of making it more pleasant to read. It is interesting that the editor did not claim that these changes would necessarily render its contents more legible, simply that they would make its contents more attractive to its target readership. Among other changes, the journal had employed a different typeface, but in his initial announcement the editor did not specify the typeface in question.

One reader, González-Rodriguez (2011), pointed out that the journal had adopted a *palo seco* or sans serif typeface for both the headings of articles and their text. (The expression *palo seco* literally means "dry stick" in Spanish. It is used as a technical phrase by Spanish typographers, but there appears to be no counterpart expression to refer to serif typefaces. Some authors use *tipografía con remates* and *tipografía sin remates*—literally, with and without finishing—and others use *tipografía con*

3.2 Dissenting Voices

adornos and *tipografía sin adornos*—literally, with and without ornamentation—while others simply borrow the foreign terms *serif* and *sans serif*.) The reader complained that the exclusive use of *palo seco* violated the general custom in Spanish typography of using sans serif typefaces for headings but serif typefaces for the body text of articles. Specifically, he claimed that adopting a serif typeface facilitated a reader's eye movements in following a line of text. (This idea was discussed but dismissed in Sect. 1.2.) In contrast, he claimed that the adoption of a sans serif typeface meant that readers' eye movements were in danger of being lost in a "river" of white spaces.

González-Rodriguez allowed two exceptions to this generalisation. One involved the preparation of texts for younger readers; the other was the situation of reading written texts on computer screens. In both cases, he argued, sans serif typefaces provided a clearer image. He cited one study with regard to younger readers, who will be considered in Chap. 6 of this book; nevertheless, his claim is clearly not relevant to the task of experienced clinicians reading articles in a specialist journal. He cited no empirical evidence with regard to reading from computer screens, and so this is just another example of "everyone knows" discussed in Sect. 1.1. The issue of reading from screens will be the focus of Part II of this book.

In his response, the editor identified the sans serif typeface that the journal was now using as Akzidenz-Grotesk (Errando Oyonarte, 2011b). As he pointed out, this was devised as long ago as 1898 and had since been used by a wide variety of organisations and agencies; hence, it was by no means a novel and untested intrusion into the publishing world. Even so, he acknowledged that the panel which had recommended the "new look" for the journal had not included any experts on typography. He also argued that the journal should be open to the possibility that readers would increasing tend to access articles online rather than printed on paper. On González-Rodriguez's (2011) account, therefore, the adoption of a sans serif typeface should actually enhance the journal's legibility for those readers. The editor agreed to keep the situation under review in the future, although in fact at the time of writing the journal still uses the same sans serif typeface throughout. Indeed, since 2013 the journal has published articles in both Spanish and English using the same appearance, the same format, and the same sans serif typeface.

There is one example of a journal that has flip-flopped on this matter. In 2009, the journal *Brain* moved from a serif typeface to a sans serif typeface, but in 2015 it moved back to a serif typeface; on the latter occasion, the editor made the comment: "Whether serif or sans-serif typefaces are more readable has been addressed sporadically with psychophysical studies, without a clear conclusion" (Kullmann, 2015, p. 1). Whether this statement provides an accurate account of the relevant research literature is the focus of Part I of this book.

3.3 Are Serifs Purely Decorative?

One simple idea can be rejected at the outset. This is the assumption that serifs are purely decorative and superfluous to the task of identifying individual letters (e.g., Arditi & Cho, 2005; Burt, 1959, p. 8). There is good experimental evidence that readers identify the specific typefaces that they are reading before they identify the individual letters or words presented in those typefaces. For instance, words take longer to identify if their constituent letters are printed in different typefaces than if they are printed in one single typeface. This is true, in particular, if the letters are printed in both serif and sans serif typefaces (Adams, 1979; Krulee & Novy, 1986; Sanocki, 1987, 1988). Neurophysiological research indicates that typeface-specific information is processed within the right hemisphere of the brain, whereas typeface-independent information is then processed within the left hemisphere of the brain (Schweinberger et al., 2006; Vaidya et al., 1998).

Weaver (2014) described the case of a 52-year-old woman with a history of complex epileptic seizures that in recent years had been triggered specifically by reading. She herself had observed that her seizures were associated with reading material printed in serif typefaces (such as Palatino or Times New Roman) but not with reading material printed in sans serif typefaces (such as Arial or Verdana), although she also had intermittent spontaneous seizures. Her observation was confirmed electrophysiologically by asking her to read the first three pages of Charles Dickens' novel, *A Tale of Two Cities*, typed in Times New Roman or Arial typefaces. Weaver suggested that the presence of serifs constituted more complex visual information that led to the activation of the hyperexcitable neuronal network responsible for her seizures.

The neurological condition of synaesthesia, in which perceiving an object in one mode stimulates the perception of a quite different mode, provides another example. A common form is grapheme–colour synaesthesia, in which specific characters, while printed in black, are seen as coloured. Weaver and Hawco (2015) described a patient who tended to perceive the letters *ll* (as in *silly*) in a vivid blue colour. The effect was more vivid for words presented in serif typefaces than for words presented in sans serif typefaces. However, if the *ll* was printed in a sans serif typeface but the rest of the word was printed in a serif typeface, the effect was more vivid than if the *ll* was printed in a serif typeface but the rest of the word was printed in a sans serif typeface. These findings show that serifs have a functional role that is not superfluous to letter recognition. This objection also applies to the idea that serifs only serve as visual noise (or as "cluttering incoming visual information", as was suggested by Woods et al., 2005, p. 97).

3.4 Conclusions

This chapter introduced Part I of this book by summarising the attitudes of 20th-century typographers, who almost without exception considered that serif typefaces were easier to read than sans serif typefaces when printed on paper. During the twenty-first century, any dissenting voices have mainly come from journal editors in medicine and bioscience, who have tended to recommend the use of sans serif typefaces for the contents of their journals but have not provided any supporting evidence. This chapter also considered but dismissed the idea that serifs are purely decorative and superfluous to the task of identifying individual letters.

Open Access This chapter is licensed under the terms of the Creative Commons Attribution 4.0 International License (http://creativecommons.org/licenses/by/4.0/), which permits use, sharing, adaptation, distribution and reproduction in any medium or format, as long as you give appropriate credit to the original author(s) and the source, provide a link to the Creative Commons license and indicate if changes were made.

The images or other third party material in this chapter are included in the chapter's Creative Commons license, unless indicated otherwise in a credit line to the material. If material is not included in the chapter's Creative Commons license and your intended use is not permitted by statutory regulation or exceeds the permitted use, you will need to obtain permission directly from the copyright holder.

Chapter 4
The Legibility of Letters and Words

4.1 Reading Letters and Words in Serif and Sans Serif Typefaces

The earliest experiments on the legibility of printed material were concerned with the relative legibility of individual letters presented in isolation in either uppercase or lowercase in the *same* serif typeface using the short-exposure method or the distance method. Other research was concerned with the relative legibility of characters presented in *different* serif typefaces. However, some researchers included one or more sans serif typefaces together with a range of serif typefaces in investigations of the legibility of letters and words:

- Griffing and Franz (1896) measured the "illumination threshold" of letters constituting from one to four words in a line. In this method, the distance between a faint light source and the material was progressively reduced until the letters could be correctly reported. They included uppercase letters in both thick and thin versions of the sans serif typeface Block.
- Roethlein (1912) used the distance method to measure the legibility of individual letters presented in the same typeface and included the sans serif typefaces Franklin Gothic and News Gothic.
- Pyke (1926) measured the legibility of both meaningful and meaningless strings of letters presented in the same typeface, including the sans serif typeface Lining Grotesque. He used a speed-of-reading test, a letter-cancellation task in which participants had to cross out all occurrences of the letters *e* and *t* in a page of nonsense material, and a task which involved reading aloud coherent text (pp. 47–58).
- Paterson and Tinker (1932) used a speed-of-reading test in which the participants had to read short passages and in each case to identify a word that conflicted with the passage's meaning. They used a number of different typefaces including the sans serif typeface Kabel Light.

- Webster and Tinker (1935) employed the distance method to measure the legibility of individual words in different typefaces and also included Kabel Light.
- Luckiesh and Moss (1937) measured the visibility threshold of individual lowercase letters and included a sans serif typeface in light, medium, and bold font. They did not identify this typeface, but Lund (1999, p. 116) suggested that it was Kabel.
- Luckiesh and Moss (1942, pp. 159–162) carried out a similar experiment and included the sans serif typeface Metrolite No. 2.

None of these researchers focused upon the difference between serif and sans serif typefaces, but the results that they presented indicate that the legibility of sans serif typefaces was not markedly different from the legibility of serif typefaces that were in common use at the time.

Ovink (1938) carried out two experiments to investigate the typographical factors that might influence the legibility of printed letters. His first experiment used the short-exposure method, and he presented individual lowercase letters in isolation in both serif and sans serif typefaces (pp. 23–37). The second experiment used a version of the distance method in which the printed material was presented in a fixed location and the participants approached it in gradual steps until it could be perceived and reported correctly. The material consisted of individual uppercase and lowercase letters presented in isolation in one of several sans serif typefaces or in one of two serif typefaces (Lo and Poster Bodoni) (pp. 38–71). The results that Ovink obtained using both methods show that the legibility of the sans serif typefaces was not markedly different from that of the serif typefaces.

Korean is another language where the alphabet can be rendered in either serif (or Ming) typefaces or sans serif (or Gothic) typefaces. Two studies have compared the legibility of the two kinds of typeface, but they yielded contradictory findings. Kong et al. (2011) asked ten older and ten younger adults to read aloud sets of four one- or two-syllable letters of varying sizes and then to rate how much discomfort they had experienced when reading each set on a 4-point scale. The sets of letters were presented either on paper or on a computer screen using either an unspecified Ming typeface or an unspecified Gothic typeface. When the letters were presented on paper, the participants' reading speed was faster and their discomfort was less with the sans serif typefaces than with the serif typefaces. Nevertheless, the relevant differences were small in magnitude and unlikely to be of practical importance. The results were similar in both age groups.

Kim et al. (2015) presented 14 Korean students with pairs of two-syllable words printed side by side in 26-point type. On each trial, one member of the pair had been designated as the target, while the other was a distractor, and the participants' task was to read aloud the target in each pair. The pairs of letters were presented either on paper or on the screen of a smartphone in one of two serif typefaces (Batung or Gungseo) or in one of two sans serif typefaces (Dodum or Gulim). Afterwards, the students were asked to rate the typefaces of the stimuli in terms of their ease of reading, their familiarity, and their comfort. When the words were presented on paper, the participants' reading time was significantly faster for the serif typefaces than for the sans serif typefaces. Despite the pattern of results for their response times, they gave the highest ratings to the sans serif typeface Dodum and the lowest ratings to the serif typeface Gungseo.

Wilkins et al. (1996) had devised the Rate of Reading Test for children with reading difficulties. The child was presented with a display of 10 lines, each consisting of a random ordering of the same 15 common words. For instance, the first line of one such display was "come see the play look up is cat not my and dog for you to", except that the spacing between successive words was only 0.36 mm. This made the display resemble horizontal stripes and thus rendered it visually stressful. (Horizontal stripes are known to induce eye strain, visual illusions, headache, and—in people with photosensitive epilepsy—seizures: A. Wilkins et al., 1984.) The researchers argued that the use of random ordering minimised the linguistic and semantic aspects of reading that tended to be emphasised in more conventional reading tests. Children were timed while they read aloud the words in each display and were scored on the number of words that they had read correctly per minute. The original version of the Rate of Reading Test only used the serif typeface Times. However, Svensson (2019) developed a Swedish version of the test and administered it to 45 adults aged between 22 and 83 years. The test was administered twice in Times New Roman and twice in Times Sans Serif, a sans serif variant designed by Mundo da Lua. The average reading speed was 168 words/min in both conditions, and the difference between them was not statistically significant ($p = 0.54$).

4.2 The "Stripiness" of Printed Words

Wilkins et al. (2007) suggested that the legibility of letters or words might depend upon their shape and, in particular, upon the extent to which letters' vertical strokes were relatively evenly spaced, a phenomenon that typographers refer to as their *rhythm* but which Wilkins et al. referred to less formally as their "stripiness" (i.e., the extent to which an image of a word approximated a pattern of vertical stripes). They suggested that this could be measured by the height of the first peak of the autocorrelation between an image of a word and a second, horizontally displaced image of the same word. They explained this measure by asking readers to imagine two identical transparencies containing a single word placed on top of one another on an overhead projector.

> When the transparencies are in register [i.e., exactly in line], a maximum amount of light will be transmitted through the combined transparencies. . . . If the top transparency is moved horizontally across the bottom transparency, the amount of light transmitted is initially reduced because the letter strokes in one version of the word block the spaces in the other version. As the displacement continues, however, and neighbouring letter strokes come into register, so the amount of light transmitted increases. As the top transparency is displaced still further, the amount of light transmitted once again decreases and then increases again. The light transmitted varies with horizontal position according to a function with peaks and troughs. This function is, in effect, the horizontal autocorrelation. (pp. 1788–1789).

As examples, Wilkins et al. gave the words "mum" and "over". (In academic texts, these words would normally be rendered in an italic font. On this occasion, I have presented the words in a regular font with inverted commas to help readers to appreciate the differences in the words' shape.) The former has fairly evenly spaced vertical strokes, high periodicity, and a relatively high first peak (high stripiness), but the latter has very few vertical line elements, low periodicity, and a relatively low first peak (low stripiness).

Wilkins et al. asked ten students to rate each of 40 common words printed in the serif typeface Times New Roman in terms of their stripiness on a scale from 0 (not at all stripy) to 10 (very stripy). They found that their mean rating for each word was highly correlated with the first peak in its horizontal autocorrelation ($r = 0.688$). In short, "words with a high first peak in the autocorrelation were rated as having a striped appearance" (p. 1791). Wilkins et al. then asked 32 university students and staff to read aloud 22 common monosyllabic words. The words were divided into those with high and low first peaks and were printed either in a single column or as a random paragraph of 18 lines in either Times New Roman or the sans serif typeface Arial. There was a large effect of autocorrelation, such that words with a high first peak were read more slowly than were words with a low first peak. Wilkins et al. confirmed this finding in two experiments using words with no ascenders or descenders that were printed in either Times New Roman or the sans serif typeface Geneva. (This indicated that the difference in reading speed was not due to the presence or absence of ascenders or descenders.) They also confirmed this finding in two experiments where participants silently scanned passages of randomly ordered words with the aim of finding pairs of target words.

In addition, Wilkins et al. compared the first peak in the horizontal autocorrelation of 1,000 words printed in different typefaces of similar x-heights. The value of the first peak in the Times New Roman was very highly correlated with its value in the serif typeface Palatino ($r = 0.95$), but it was less highly correlated with its value in the sans serif typeface Arial ($r = 0.68$). The first peak tended to be highest in Times New Roman, somewhat less in the sans serif typeface Lucida Sans, and lowest in the serif typeface Palatino and the sans serif typeface Arial, although the differences were small in magnitude. Wilkins et al. used the same corpus of 1,000 words to compare the horizontal autocorrelation in the serif typefaces Times New Roman and Tahoma, the sans serif typefaces Arial and Verdana, and the slanting sans serif typeface Sassoon Primary (discussed in Sect. 7.4). They did not report the detailed findings, but Verdana had the lowest first peak.

In two of their experiments, Wilkins et al. directly compared the reading times for different typefaces. Random paragraphs were read significantly more quickly in the sans serif typeface Geneva than in the serif typeface Times New Roman. However, there was no significant difference between the reading speeds in Times New Roman and the sans serif typeface Arial, regardless of whether the words were presented in a single column or in random paragraphs. In short:

- Different words printed in the same typeface vary in the first peak of their horizontal autocorrelation (or vertical stripiness).
- The same words printed in different typefaces vary in the first peak of their horizontal autocorrelation.

- Words of low vertical stripiness are read more quickly than are words of high vertical stripiness.

However, the results obtained by Wilkins et al. leave it uncertain whether these phenomena lead to variations in how quickly words in different typefaces are read. Subsequently, Wilkins and his colleagues carried out further research using words presented on computer monitors, and this is described in Sect. 11.2.

4.3 Confusions Among Letters in Serif and Sans Serif Typefaces

Many early studies found that errors in the tachistoscopic recognition of individual letters and words were the result of confusions among visually similar letters (for a review, see Vernon, 1931, pp. 114–120, 145–150, 158–159). In the light of such evidence, Legros (1922, p. 11) claimed that serifs made letters easier to discriminate and identify. However, Vernon (1929) noted that other studies had found that, in reading connected text, words tended to be perceived in a holistic manner rather than letter-by-letter, and hence confusions among individual letters should be much less important. She presented adults with different kinds of material using a tachistoscope. She found that the proportion of errors based upon similarity of *appearance* declined from 82% for groups of unrelated words to 14% for longer sentences, whereas the proportion of errors based upon similarity of *meaning* increased from 2% for groups of unrelated words to 57% for longer sentences.

Vernon argued that, "when the meaning of the material read was fully comprehended, typographical errors were few in tachistoscopic reading, and would be negligible in normal reading" (p. 35). Elsewhere, Vernon (1931, pp. 171–172) concluded that young children beginning to read might be liable to confuse visually similar letters but that this was of much less importance in normal adults' reading. An implication of this is that, even if the presence or absence of serifs influences the discrimination or identification of individual letters, it should have little or no impact upon the reading of connected text by literate adults.

Tinker (1963, p. 36) argued that in practice serifs might serve either to enhance or impair the relative differentiation of individual lowercase letters. Harris (1973) presented individual lowercase letters tachistoscopically either to the left or to the right of the point of fixation in the sans serif typefaces Gill Medium and Univers Medium or in the serif typeface Baskerville. He found that different letters were more likely to be confused when presented in Baskerville. He suggested that serifs on letters with a single vertical stroke (such as i, j, and l) rendered them more distinctive and hence less likely to be confused with one another. On the other hand, he also suggested that serifs on letters with more than one vertical stroke (such as h, n, and u) rendered them less distinctive and hence more likely to be confused with one another. Beier and Dyson (2014) obtained analogous results using artificial typefaces with a version of the distance method. However, Vernon's (1929) findings would imply that such confusions would be much less likely if letters were presented in the context of meaningful text, as in normal reading.

4.4 Measuring Visual Acuity

Some of the earliest charts for measuring visual acuity were developed by Snellen (1862) (see Fig. 1.4 in Sect. 1.3). These contained rows of uppercase letters and single digits based on a 5 × 5 grid; this yielded a slab serif style akin to a typeface that was then known as Egyptian Paragon, in which the width of the main strokes and the width of the serifs were one fifth the height of a letter. (Both were the size of the cells in the 5 × 5 grid.) Successive rows contained increasing numbers of symbols of decreasing size, and visual acuity was scored according to the smallest row that could be read accurately in each eye.

Over the next century, other researchers developed versions of these charts using different layouts and sequences of letters; some followed Snellen in using a slab serif style, whereas others adopted a sans serif style (see Bennett, 1965, for a review). Cowan (1928) asserted that "Gothic" (sans serif) letters were more easily distinguished than "block" (slab serif) letters (p. 290), but he provided no reference or any other source for this assertion. Hetherington (1954) tested the visual acuity of 100 boys aged 8–17 using an unspecified version of a Snellen chart; he noted that different letters of the same size varied in their legibility, but he concluded that the boys' errors were generally the result of confusions among visually similar letters.

In fact, since the 1950s, charts with sans serif styles of lettering have been widely adopted for measuring visual acuity. Examples include those devised by Sloan (1959), the British Standards Institute (1968), Deederer (1968, 1970), and Bailey and Lovie (1976). The British Standards Institute (1968) commented that this development in measures of visual acuity was "in keeping with modern trends in typography" (p. ii), while Deederer (1968) remarked that it was appropriate for testing drivers, since they would frequently encounter sans serif lettering on traffic signs.

Richards (1965) compared the visual acuity of 103 volunteers who were tested using Sloan's sans serif chart and a Snellen chart containing lettering with slab serifs; people who had slight uncorrected astigmatism found the letters with slab serifs more confusing under conditions of low luminance, but otherwise there was very little overall difference between the results obtained using the two styles. Richards (1978) subsequently replicated these findings with a sample of 175 volunteers stratified by age between 16–25 years and 66–75 years (that is, the sample contained 25 volunteers from each decade of the adult life span). Bailey and Lovie's (1976) instrument is known as the LogMAR chart (the acronym stands for Logarithm of the Minimum Angle of Resolution), and nowadays it is generally regarded as the most accurate measure of visual acuity.

4.5 Conclusions

The earliest research on the legibility of different typefaces was concerned with recognising individual letters and words under different conditions. The vertical "stripiness" of individual words can be defined in terms of their horizontal autocorrelation. This seems to affect how quickly they can be read, but it is unclear whether this leads to differences among typefaces. There is a separate line of research concerned with evaluating visual acuity, going back to the construction of optical charts in the middle of the nineteenth century. In both fields of research, the most common finding—the *modal* finding—is that there are no differences in the legibility of letters and words printed in serif and sans serif typefaces. Confusions among visually similar letters were originally considered to be a primary determinant of legibility, but these appear to be less important when skilled readers are presented with meaningful text.

Open Access This chapter is licensed under the terms of the Creative Commons Attribution 4.0 International License (http://creativecommons.org/licenses/by/4.0/), which permits use, sharing, adaptation, distribution and reproduction in any medium or format, as long as you give appropriate credit to the original author(s) and the source, provide a link to the Creative Commons license and indicate if changes were made.

The images or other third party material in this chapter are included in the chapter's Creative Commons license, unless indicated otherwise in a credit line to the material. If material is not included in the chapter's Creative Commons license and your intended use is not permitted by statutory regulation or exceeds the permitted use, you will need to obtain permission directly from the copyright holder.

Chapter 5
Reading and Comprehending Text

5.1 Reading Text in Serif and Sans Serif Typefaces

Ovink (1938) included a sans serif typeface (Futura) as well as a variety of serif typefaces in experiments where participants read lines of text presented for a limited exposure (pp. 84–88) or where their eye movements were monitored while they were reading passages held in their hands in clear view (pp. 88–100). In the latter case, Ovink used a mechanical apparatus in which a rubber pad rested on one of the participants' upper eyelids. He also included a sans serif typeface (Gill Sans) as well as three serif typefaces in a further study where the short-exposure method was used to present material in the form of dictionary entries and where the participants subsequently rated the clarity of the typefaces that had been used (pp. 100–106). There were no marked differences between their responses to the sans serif typefaces and their responses to the serif typefaces.

Wendt (1969, 1994) carried out an investigation to compare the effects of typographic factors, including the legibility of a serif typeface (Bodoni) and a sans serif typeface (Futura). He constructed 16 passages in German and presented each in five columns across a sheet of paper. He asked roughly 2,000 students to read one of the passages and recorded the number of words read in 3 min. There was a slight mean advantage of 8.62 words for passages in the Futura typeface, but this did not achieve statistical significance. Wendt argued that modern readers were equally familiar with both styles of typeface. Indeed, in a subsequent survey of Australian students, the serif typeface Press Roman was ranked only marginally higher in their overall preference than the sans serif typeface Univers (Bell & Sullivan, 1981).

Taylor (1990) arranged for the instructors of four remedial reading classes at a US high school to administer reading tests to groups of students aged 15–16. Over three weeks, they were timed reading excerpts from their reading workbooks, one printed in the sans serif typeface Helvetica, the other printed in the serif typeface Times Roman. (As was mentioned in Sect. 1.2, this is a similar typeface to Times New Roman that was developed by a rival printing company; it is sometimes known just as "Times".) Across 74 students, the median difference in their reading rate scores on the two typefaces was zero (p. 50). Over the next two weeks, they were given sheets of paper containing other excerpts printed in both of the typefaces, and they were asked to choose one version to read aloud. There was no significant difference in their preferences in either week.

A major research issue is that actual examples of serif and sans serif typefaces tend to differ on a variety of other characteristics (Beier & Larson, 2010). Arditi (2004) had devised software to generate typefaces that differed only in the presence or absence of slab serifs and in the size of the serifs. Arditi and Cho (2005) used this tool to construct lowercase typefaces of uniform thickness with slab serifs that extended for 0% (sans serif), 5%, or 10% of their cap height. Two individuals with normal vision and two with impaired vision were asked to read aloud three "passages" in which 400 words had been randomly ordered to yield "scrambled" text. Arditi and Cho calculated each participant's "reading speed" by dividing the number of characters in the correctly read words by the time taken to read each passage. The participants with normal vision obtained higher scores than those with impaired vision, but there was no significant variation in their reading speed as a function of serif size and a fortiori no significant effect of the presence or absence of the slab serifs. As Arditi and Cho noted, the small number of participants was a major limitation of their study.

5.2 Comprehending Text in Serif and Sans Serif Typefaces

In Sect. 2.2, it was noted that asking participants to read continuous meaningful text provides less opportunity for researchers to impose experimental control over their reading behaviour. To address this issue, some researchers have focused on their participants' comprehension of such material rather than upon its legibility per se. This could be justified on the grounds that, in order to be comprehended, written material must first be read; thus the legibility of a text places an upper limit on how much of it can be comprehended. A different argument was put forward by Gasser et al. (2005). They suggested that, insofar as a message was easy to read, fewer attentional resources would need to be devoted to reading the message, leaving more resources available for the processing of the information that it contained. However, Reynolds (1979) argued that the value of comprehension tests in the measurement of legibility was questionable, insofar as comprehension implicated cognitive skills of a higher order than those that were required for the accurate perception of the text. Wilkins et al. (1996) also claimed that conventional reading tests tended to emphasise the linguistic and semantic aspects of reading rather than the purely visual aspects (see Sect. 4.1). Thus, factors affecting comprehension might not be relevant to measuring legibility.

5.2 Comprehending Text in Serif and Sans Serif Typefaces

Fox (1963) compared two typefaces intended for use on typewriters: Standard Elite, a conventional serif typeface; and Gothic Elite, a sans serif typeface in which lowercase letters are replaced by small capitals. Both typefaces are monospaced or non-proportional (each character occupies the same width), and small capitals in Gothic Elite occupy the same width as lowercase letters in Standard Elite. The participants read two passages, one in each typeface, silently but as quickly as possible. After reading each passage, they were required to recount the story that it contained, and their comprehension was rated as "good" or "poor". There was no significant difference in the time taken to read passages in the two typefaces or in their comprehension of their content.

Poulton (1965) asked 375 adult volunteers to read two passages of about 450 words in the same typeface. They were allowed 90s to read each passage and were given a test of their comprehension of ten key points from the passage. The passages had been printed in seven different typefaces: three serif typefaces (Baskerville, Bembo, and Modern) and four sans serif typefaces (Gill Medium, Grotesque 215, and two versions of Univers). On the first passage, Poulton found no significant differences among the comprehension scores, which he ascribed to the participants becoming familiar with the general procedure and the particular typeface that they had to read. On the second, there was significant variation among the sans serif typefaces but not among the serif typefaces; more important, none of the serif typefaces yielded scores that were significantly different from those of any of the sans serif typefaces.

A fundamental question using this paradigm is whether a test on the content of a passage administered immediately after its presentation is a test of comprehension or simply a test of factual recall or verbatim memory (Hartley et al., 1975). Poulton and Brown (1967) used the same procedure as in Poulton's (1965) study, but they remarked that their measure of "comprehension" was "more correctly described as a measure of memory" (p. 219). They found that requiring the participants to read a passage aloud led to poorer performance on the early key points in the passage but to better performance on the last key point in comparison with requiring the participants to read the passage silently. Unfortunately, Poulton and Brown had not matched the key points and their associated questions for difficulty, and consequently the theoretical interpretation of these results remains unclear.

Soleimani and Mohammadi (2012) evaluated different typefaces in Iranian students who had been selected for having an intermediate proficiency in English. This included a reading comprehension test based on a passage from a widely used English-language textbook: it was presented in the sans serif typeface Arial for 42 students and in in the serif typeface Bookman Old Style for 47 students. There was no significant difference between the two groups in their reading speed, in an immediate test of their reading comprehension, or in a multiple-choice test of their memory for ten key points from the passage that was administered 2 weeks later.

Serif and sans serif typefaces are also used in some non-Western alphabets. Akhmadeeva et al. (2012) asked 238 Russian medical students to read a passage about the history of neurology in Russia. The passage had been printed in Cyrillic script using ParaType, a family of artificial typefaces: 108 students were shown the passage in a serif typeface, and 130 students were shown the passage in a sans serif typeface with the same x-height. The students were given 1 min to read the passage and were then asked ten multiple-choice questions about its content. Akhmadeeva et al. found no sign of any difference either in the mean number of words that the two groups had read or in the mean number of questions that they had answered correctly.

5.3 The Connotative Meaning of Typefaces

Some researchers argued that typefaces could serve as carriers of *connotative meaning* (reflecting their associations with different attitudes, experiences, and emotions) as well as carriers of *denotative meaning* (reflecting factual information). Subjective impressions of the legibility of different typefaces can be regarded as just one aspect of their connotative meaning. (German-speaking writers sometimes refer to this quality as their *Atmosphärenwert* or "atmosphere value". North American writers sometimes talk about the "personality" of different typefaces.) This aspect might in principle affect their legibility for different readers. In this kind of research, participants are asked to report on their experiences and preferences when reading material printed in different typefaces. Once again, examples can be presented either individually or in groups of two or more for comparison, and the self-reports can be collected either informally (for instance, through interviews) or more formally (for instance, through the use of rankings or rating scales).

As an example of a formal approach, Tinker and Paterson (1942) asked different groups of participants to arrange samples of ten different typefaces in order from the most legible to the least legible and from the most pleasing to the least pleasing. They found that their judgements of legibility and pleasantness demonstrated "remarkable agreement" (p. 40), to the extent that they could be regarded as being equivalent to one another. As Dreyfus (1985) pointed out, readers' preferences may be irrelevant if they have no choice in whether or not to read something (such as an airline schedule or a railway timetable) but may be crucial if they can choose what they read (as in product information or voting literature).

Another example of a formal approach is the semantic differential. This was devised by Osgood et al. (1957) to measure participants' attitudes to objects and concepts. The participants provide evaluations of these using bipolar rating scales; typically, these are 7-point scales in which the middle category is neutral between the two poles. Their responses are subjected to factor analysis to yield higher-order dimensions that are regarded as reflecting underlying aspects of connotative meaning. Research studies in a wide variety of domains and cultures converged on three overarching dimensions that reflected variations in peoples' attitudes: evaluation (good vs. bad), potency (strong vs. weak), and activity (active vs. passive) (Osgood et al., 1975). Hofstätter (1966) devised a similar methodology for German-speaking countries, which he described as a *Polaritätsprofile* (polarity profile).

5.4 Connotations of Serif and Sans Serif Typefaces

Connotative meaning is potentially important in the world of advertising. Berliner (1920) initiated this line of inquiry by asking students to rank order different hand-lettered styles on different dimensions for advertising different products. She found that their rankings of appropriateness were different for different products. Subsequent researchers confirmed this when using actual typefaces (Davis & Smith, 1933; Poffenberger & Franken, 1923; Schiller, 1935). They found no clear difference between the ratings given to serif and sans serif typefaces; other features seemed to be more important in determining the appropriateness of different typefaces to different products. Ovink (1938, pp. 127–177) noted that these studies had only used typefaces that were in common use in advertising displays in the United States. He chose 17 typefaces, including some that were more widely used in Europe. He asked 68 participants to rate the appropriateness of each of the typefaces for advertising purposes on eight dimensions. The 17 typefaces varied on most dimensions, but there was no clear difference overall between serif typefaces and sans serif typefaces.

Using the semantic differential methodology developed by Osgood et al. (1957), Tannenbaum et al. (1964) presented participants with the English alphabet printed in both uppercase and lowercase in both upright and italic fonts in two serif typefaces, Bodoni and Garamond, and two sans serif typefaces, Spartan and Kabel. A total of 75 participants rated each display on 25 scales. There were no significant differences between the ratings given to the serif typefaces and those given to the sans serif typefaces.

Wendt (1968) asked 70 participants to rate 35 typefaces using an adapted version of Hofstätter's (1966) semantic differential. Eleven were sans serif typefaces (seven variants of Folio and four variants of Futura). Each typeface was presented on a printed card containing the alphabet in lowercase, the alphabet in uppercase, and the ten single digits. Each was evaluated by ten participants on 63 7-point rating scales. Factor analysis was used to reduce their ratings to four broad dimensions, but there were no clear differences between the serif and the sans serif typefaces on these dimensions. Cluster analysis of the ratings yielded three clusters, but each of these contained both serif and sans serif typefaces. In other words, the participants' ratings differentiated among the 35 typefaces, but they did not differentiate systematically between the serif typefaces and the sans serif typefaces.

Benton (1979; Rowe, 1982) asked 24 participants to evaluate ten typefaces on 26 bipolar 7-point scales. Five were general typefaces in general use, including one sans serif typeface (Helvetica); five were novelty typefaces. Each was presented on a printed sheet containing the alphabet in lowercase, the alphabet in uppercase, the ten single digits, and common punctuation marks. Factor analysis was used to reduce their ratings to five broad dimensions, which Benton labelled "potency", "elegance", "novelty", "antiquity", and "evaluation", and which she used to calculate scale scores for the ten typefaces. The serif typefaces (Bodoni, Garamond, Palatino, and Times Roman) did not differ significantly from each other on any of the five dimensions.

Moreover, Helvetica only differed from these typefaces on antiquity, where it was seen as being relatively modern.

Bartram (1982) conducted a similar study in which 38 design students and 52 students of other disciplines were asked to rate 12 typefaces on 18 bipolar scales. A factor analysis of the ratings provided by the design students yielded four dimensions: the three hypothesised by Osgood et al. (1957) (evaluation, potency, and activity) and a fourth dimension concerned with mood. Bartram then compared the scores on these dimensions given to the 12 typefaces by the design students and the other students. They included three regular upright typefaces: the serif typeface Times New Roman and the sans serif typefaces Futura Medium and Univers 67. Bartram did not compare the ratings of these typefaces directly, but their profiles were relatively similar.

Morrison (1986) obtained ratings of four serif typefaces (Egyptian, Modern, Old Style, and Transitional) and one sans serif typeface (Contemporary) on 25 bipolar dimensions taken from Tannenbaum et al. (1964). There were three groups, each of 14 participants: typography students, technology students, and students of other subjects. They were presented with "text" consisting of statistical approximations to English in all five typefaces in three weights and in both upright and italic font. There were no significant differences among the three groups in terms of their mean ratings on seven scales that included the three primary factors identified by Osgood et al. (1957). The only significant difference between the serif typefaces and the sans serif typeface was that the latter received slightly higher ratings than the former on the potency factor, which Morrison attributed to sans serif typefaces being used on public signs that had an association with authority (such as freeway and airport signage).

Tantillo et al. (1995) asked 250 students to evaluate examples of six typefaces on 28 bipolar 7-point scales. There were three serif typefaces (Century Schoolbook, Goudy Old Style, and Times New Roman) and three sans serif typefaces (Avant Garde Gothic, Helvetica, and Univers). Significant differences emerged on 26 scales: "The serif type styles... are rated as more elegant, charming, emotional, distinct, beautiful, interesting, extraordinary, rich, happy, valuable, new, gentle, young, calm, and less traditional than the sans serif type styles. Serif styles have more personality, freshness, high quality, vitality, and legibility, but the sans serif group is more manly, powerful, smart, upper-class, readable, and louder than the serif styles" (p. 452).

These results are anomalous when compared with the findings of earlier research. A procedural difference is that Tantillo et al. only presented the nonsense word "NRESTA" in uppercase as the example of each typeface to be evaluated, but other studies had used longer examples involving both uppercase and lowercase letters. The rating forms that Tantillo et al. employed often showed the more positive pole on the left end of each scale, but it sometimes appeared on the right end, which may have led to a degree of confusion on the students' part. This might explain two curious findings: first, sans serif typefaces are usually regarded as being more modern in their appearance than serif typefaces, yet Tantillo et al.'s participants rated serif typefaces as being younger and less traditional; second, the sans serif typefaces were rated as being significantly more *readable* but as significantly less *legible* than the serif typefaces.

5.5 Conclusions

A number of studies have evaluated the role of typographic variables (including the presence or absence of serifs) in reading continuous text. Asking participants to read continuous text allows less scope for experimental control, and so other researchers have instead focused on participants' comprehension of written material. In both cases, the modal finding is that there are no significant differences between text printed in serif typefaces and text printed in sans serif typefaces. Subjective impressions of the legibility of different typefaces can be regarded as one aspect of their connotative meaning, and other researchers have asked participants to evaluate typefaces on different dimensions using single rating scales or semantic differentials. The modal finding is that there are no significant differences in readers' overall preference between serif and sans serif typefaces, nor any significant differences in the connotations of serif and sans serif typefaces.

Open Access This chapter is licensed under the terms of the Creative Commons Attribution 4.0 International License (http://creativecommons.org/licenses/by/4.0/), which permits use, sharing, adaptation, distribution and reproduction in any medium or format, as long as you give appropriate credit to the original author(s) and the source, provide a link to the Creative Commons license and indicate if changes were made.

The images or other third party material in this chapter are included in the chapter's Creative Commons license, unless indicated otherwise in a credit line to the material. If material is not included in the chapter's Creative Commons license and your intended use is not permitted by statutory regulation or exceeds the permitted use, you will need to obtain permission directly from the copyright holder.

Chapter 6
Reading in Context

6.1 The Importance of Context

Whittemore (1948) argued that the legibility of different typefaces depended upon the context in which they were used. Readers may develop expectations with regard to the kinds of context in which particular typefaces are appropriate. A problem with much of the research described thus far is that it did not provide readers with any sensible context for their reading (Schriver, 1997, p. 277). Some researchers have endeavoured to address this issue.

Zachrisson (1965, pp. 156–162) investigated the attitudes of experts and non-experts to whether material was printed in serif or sans serif typefaces. Typography experts and students from various subject areas were shown samples for each of six themes. A serif typeface was preferred for a wedding invitation, a perfume advertisement, and the title page for a book of lyrical verse, but a sans serif typeface was preferred for an invitation to an art exhibition, the title page for a book on modern architecture, and an advertisement for an oil stove. The rankings given by the experts and the non-experts were relatively similar.

Hvistendahl and Kahl (1975) prepared four newspaper stories of 250–400 words in four serif typefaces (Imperial, News #2, News Bold, and Royal) and four sans serif typefaces (Futura, Helvetica, News Sans, and Sans Heavy). Each of 200 subjects was asked to read at their normal reading pace two stories in serif typefaces and two stories in sans serif typefaces; in each case, one story was printed in 10.5-point type, while the other was printed in 14-point type. For two stories, the serif typefaces yielded a significantly faster mean reading time than the sans serif typefaces; for the other two stories, there was no significant difference in the reading times. Hvistendahl and Kahl then showed each participant two out of eight stories set in *both* a serif typeface *and* a sans serif typeface; in each case, one story was printed in 10.5-point type, and the other was printed in 14-point type. The participants were asked to express their preference between the two typefaces in which each story was printed. Overall, the serif typefaces were preferred 68% of the time, while the sans serif typefaces were preferred only 32% of the time. Nevertheless, it should be noted that Hvistendahl

© The Author(s) 2022
J. T. E. Richardson, *The Legibility of Serif and Sans Serif Typefaces*,
SpringerBriefs in Education, https://doi.org/10.1007/978-3-030-90984-0_6

and Kahl did not ask their participants to compare stories set in serif and sans serif typefaces of the *same* size.

Moriarty and Scheiner (1984) asked 260 college students to read a page of text from a sales brochure for stereo speakers. Half the students read the text in a serif typeface (Times Roman), and half read it in a sans serif typeface (Helvetica). Independent of this, half the students read the text in regular spacing, and half read it with an 18% reduction in spacing. They were given 105 s to read the text and marked the last word that they had read when the time limit was reached. The students given a close-set type read significantly more than those given the regular type. However, there was no significant difference between the students who read the text in a serif typeface and those who read it in a sans serif typeface and no significant interaction between the two variables. Moriarty and Scheiner concluded that there was no difference in reading speed between the serif and sans serif typefaces in their study.

6.2 Serif and Sans Serif Typefaces in Newspaper Headlines

The headlines placed above articles in the body of a newspaper are usually presented in a large typeface, often in a bold font, and they may extend over two or more lines. They may be presented (a) all in capitals, (b) with initial capitals for the principal words ("title case"), or (c) with initial capitals only for the first word and any proper nouns (sometimes known as "sentence case"). Research in the first half of the twentieth century showed that text presented in lowercase was read more quickly than text presented in uppercase and more specifically that headlines in title case were read more quickly than headlines all in capitals (see Tinker, 1963, pp. 186–190). Newspaper headlines are sometimes presented in sans serif typefaces. Arnold (1956) argued that "Sans Serif... is highly readable and, more than any other typographic device, conveys an impression of a newspaper that is alert and up to date" (p. 19).

English (1944) presented three-line headlines in title case tachistoscopically for 450 ms to 45 students of journalism and psychology. He used a serif typeface (Bodoni bold), a slab serif typeface (Karnak bold), and a sans serif typeface (Tempo medium) in three sizes, and he used the number of words reported correctly as a measure of performance. Headlines presented in Bodoni and Tempo yielded significantly better performance than those presented in Karnak but were not significantly different from each other. Variations in type size had no effect upon the participants' performance. Another group of 50 students was shown pairs of headlines in different typefaces and asked to choose which member of the pair seemed easier to read. There were no significant differences among their preferences for the different typefaces.

Haskins (1958) presented 300 participants recruited through the *Saturday Evening Post* with ten different magazine articles. The subtitles and text of the articles were presented in their original form, but their headlines were presented to different participants in one of ten different typefaces, including two sans serif typefaces (Futura Light and Futura Bold). The participants were asked to judge how appropriate each headline was for the article to which it was attached using a 6-point scale. The variation in their ratings across the ten typefaces was highly significant; in particular, Futura Light was on average judged to be fairly appropriate, whereas Futura Bold was judged on average to be very appropriate for eight of the articles. These results imply that sans serif typefaces are at least as appropriate as serif typefaces for the headlines of such articles.

Click and Stempel (1968) carried out an experiment in which college students rated six newspaper front pages on 20 semantic differential scales (see Sect. 5.3). However, they had carried out a pilot study in which six front pages had used sans serif typefaces in their headlines and four had used serif typefaces. They found that sans serif typefaces and serif typefaces yielded similar ratings if they were used with similar front-page formats. As they noted, "The main source of variation in response was the format and not the typeface" (p. 130). Because of this, they only used sans serif headlines in their main experiment.

Haskins and Flynne (1974) investigated whether the choice of different typefaces for newspaper headlines affected readers' interest in the accompanying stories. They carried out interviews with 150 female heads of household, of whom 100 were asked to read through a genuine local newspaper in which a mock women's page had been inserted. This contained five articles drawn from various newspapers and magazines in which the headlines had been printed either in a serif typeface judged to be relatively "feminine" (Garamond Italic) or in a sans serif typeface judged to be relatively "masculine" (Spartan Black). The remaining 50 female heads of household read the newspaper without the women's page, together with the headlines of the five articles printed on individual white cards. The participants were asked to rate the attractiveness and interest of each page of the newspaper on a scale from 0 to 100 and to rate their interest in each of the five articles. They were then shown the mock women's page printed in ten different typefaces and were asked to rate each version using 12 semantic differential scales.

Consistent with the researchers' assumptions, two typefaces often used on women's pages of newspapers, Garamond Italic and the cursive script Coronet Light, were rated more highly on several supposedly feminine characteristics, whereas Spartan Black, which was often used on sports pages, was rated more highly on several supposedly masculine characteristics. The other typefaces were rated as being relatively neutral on these characteristics. Nevertheless, there was no significant difference in the ratings of overall reading interest given to the women's pages with headlines printed in Garamond Italic and in Spartan Black. Only one of the five articles showed a significant difference in the ratings of reading interest, where the version with a headline printed in Garamond Italic was rated more highly than the version with a headline printed in Spartan Black. Haskins and Flynne concluded that, while some typefaces used in headlines were perceived as more feminine or more masculine, this had no effect on a woman's interest in reading a women's page.

In a study described in more detail in Sect. 6.3, Wheildon (1990, pp. 18–22; 2005, pp. 61–73) presented the same participants with articles containing headlines in different typefaces and evaluated their perceived legibility by asking the participants to say simply whether the headlines were easy to read or not. When the headline was presented in a lowercase serif typeface, 92% said that it was easy to read; when it was presented in a lowercase sans serif typeface, 90% said that it was easy to read. (He did not specify the exact typefaces that he had used, and he did not explain whether "lowercase" meant title case or sentence case.) Wheildon (1990, p. 22; 2005, p. 72) concluded that there was little to choose between serif and sans serif typefaces in headlines.

6.3 Wheildon's Research

Colin Wheildon (1984) carried out a study of the impact of various typographical factors on the comprehension of newspaper copy that incorporated a comparison between serif and sans serif typefaces. He prepared revised versions of his original report in 1986 and 1990, and he also incorporated his account into a book on typography and design (Wheildon, 1995). This in turn went through several editions and revisions, and the final version was published in 2005. Wheildon's research has been cited in support of the idea that serif typefaces are more legible than sans serif typefaces in continuous text (Kempson & Moore, 1994, pp. 52, 284; Schriver, 1997, p. 274). It has, however, proved to be extremely controversial (Poole, 2012).

Wheildon recruited 300 volunteers from among the inhabitants of Sydney, Australia, and he visited them in their homes on several occasions. Their educational level tended to be higher than that of the general population (79% had graduated from high school and 23% had obtained a university degree or a comparable qualification), but none was professionally involved in printing or publishing. On each visit, they were asked to read a mock newspaper article to a time limit and were then asked ten questions to test their comprehension of its content. For each visit, the participants were randomly divided into two equal subsamples who read the article in different forms, and they were classified into three groups depending on the number of questions that they had answered correctly: between seven and ten questions, good comprehension; between four and six questions, fair comprehension; and between zero and three questions, poor comprehension (Wheildon, 1990, p. 9; 2005, pp. 134–138).

At two of the visits, the bodies of the relevant articles were presented in either a serif typeface (Corona) or a sans serif typeface (Helvetica). The sequence of administration of the two typefaces was counterbalanced across different participants, and so the comprehension of the two typefaces was compared within the same individuals. The results were analysed for the 224 participants who had participated at all of the visits. When reading an article in serif typeface, comprehension was scored as good for 67%, fair for 19%, and poor for only 14%; however, when reading an article in sans serif typeface, comprehension was scored as good for only 12%, fair for 23%, and poor for 65% (Wheildon, 2005, p. 47).

Wheildon also asked the participants who had shown either poor comprehension or good comprehension "leading questions" about their attitudes to the articles and the layout of the pages. He commented that "these responses were collected for anecdotal rather than scientific value" (Wheildon, 1990, p. 9), but he felt that they helped to explain some of the objective results (Wheildon, 2005, p. 138). He summarised the comments made by the 112 participants who had read an article intended to be of direct interest that had been presented in the sans serif typeface. Many of their comments referred to their difficulty in concentrating on the reading task. However, when they were then asked to read another article presented in the serif typeface, they reported no physical difficulties (Wheildon, 1990, p. 17; 2005, p. 48).

In introducing his research, Wheildon (1990, p. 16; 2005, p. 46) had mentioned only one previous study on the legibility of serif and sans serif typefaces (Pyke, 1926), and he did not acknowledge that the sheer size of the effect that he had found linking serif typefaces to better comprehension was clearly anomalous when taken in the context of the findings of all other research carried out up to that point. Nor did he comment on the apparent disparity between these results and his findings regarding the legibility of serif and sans serif typefaces when used in newspaper headlines (described in the previous section). Instead, he argued: "The conclusion must be that body type must be set in serif type if the designer intends it to be read and understood" (Wheildon, 1990, p, 17; 2005, p. 48).

Poole (2012) argued that Wheildon's account of his research lacked key information that would enable a sceptical reader to evaluate the study. The introduction to the expanded version of Wheildon's (1990, pp. 9–10) report and an appendix to Wheildon's (2005, pp. 133–140) book do provide additional information about his research methods, but some important details are unclear. For instance, the report states that each of the articles extended over several pages (Wheildon, 1990, p. 9). However, the book states that they were designed to fit in four columns 5 cm wide and 30 cm deep on a single page, while the examples that are provided in the book indicate that some space on the page was taken up by a headline, a by-line, and two illustrations (Wheildon, 2005, pp. 34–35). Neither account mentioned either the final number of words in each of the articles or the time allowed to read the articles and to answer the comprehension questions (Wheildon, 1990, p. 9; 2005, pp. 33–48, 137–138).

There are some additional issues with Wheildon's research. First, his general account of the research methodology suggested that each participant was asked to read one article at each visit, and that comparisons were made between their comprehension of different articles at different visits (Wheildon, 2005, p. 137). Nevertheless, he also mentioned a group of 112 participants who were tested on an article of direct interest in a sans serif typeface but who were tested on "another article with a domestic theme" in a serif typeface immediately afterwards (Wheildon, 1990, p. 17; 2005, p. 48). The latter arrangement is clearly more vulnerable to transfer or carry-over effects (for instance, due to practice or fatigue) than repeated testing separated by an interval of weeks or months.

Wheildon (1990, p. 9; 2005, p. 136) had designed his materials to measure the effects of several different variables simultaneously. For example, half the mock newspaper articles were designed to be of direct or broad interest to the participants, whereas the other half were designed to be of limited or specific interest. This manipulation produced a difference of 10 percentage points in the participants' level of comprehension (Wheildon, 2005, p. 36). Other variables included the use of capital letters in the headlines, the use of colour in the headlines or in the text, the use of justified versus unjustified text, and the use of italic font in the text. Wheildon (2005, p. 136) explained this by arguing that the different manipulations were logically separate from one another, and hence there was no need to change the samples of participants to measure their effects. However, this ignores the possibility that these effects were not *empirically* separate from one another. In other words, the apparent difference in comprehension between material printed in serif and sans serif typefaces might simply have been an artefact due to a confounding of this manipulation with one or more of the other variables.

One further possibility is that of researcher bias. Wheildon's (2005, pp. 24, 103) own comments make it clear that he had always had a deep antipathy towards sans serif typefaces. He himself had interviewed and tested all the participants in their own homes (p. 137), and so it is possible that his underlying attitudes to serif and sans serif typefaces might have (either intentionally or unintentionally) influenced how they had set about their task and thus might have influenced their comprehension. This issue should have been addressed by employing assistants to test the participants who were blind (i.e., uninformed) as to the specific research hypotheses.

6.4 More Recent Research

Schriver (1997, pp. 289–303) obtained examples of texts that might be read for each of four common purposes and presented each in both a serif typeface and a sans serif typeface with similar x-heights. The four purposes or "genres" were: (a) reading to enjoy (a two-page spread from a short story, presented in Bauer Bodoni and Univers); (b) reading to assess (a one-page business letter from a bank, presented in Palatino and Futura); (c) reading to do (a two-page spread from an instruction manual, printed in Times Roman and Helvetica); and (d) reading to learn to do (a two-page spread from a manual to help people estimate their taxes, printed in Garamond Light and Optima). Schriver presented these texts to 67 volunteers using a within-subjects design that counterbalanced the order of the texts and the order of the typefaces. The participants were asked to say which version of each text they preferred and also to say why.

Similar proportions of participants chose the serif and the sans serif typefaces. More detailed examination showed that on balance participants tended to prefer the serif typefaces for the short story and the tax manual, but they tended to prefer the sans serif typefaces for the instruction manual and the business letter. Their preferences were influenced by various factors related to the rhetorical context of each text: the mood or tone of the text; the density of the text; the contrast among the parts of the text; the legibility of the text; and the quality of printing. Schriver concluded: "This study suggests that people find serif and sans serif typefaces equally pleasing but that the situation in which they are reading may lead them to prefer one style over the other" (p. 302).

6.4 More Recent Research

McCarthy and Mothersbaugh (2002) showed a fictious advertisement about Ontario to 265 business students in their regular classes. Three aspects were varied independently and at random: serif versus sans serif typefaces taken from a family of artificial typefaces; 8-point type versus 10-point type; and types with x-heights of 50% or 70% of the associated capital letters. The participants were asked to read the material to themselves for 1 min and to circle the word that they were reading when the time limit was reached. They were then asked to read a control advertisement following the same procedure and were classified into fast or slow readers using a median split on the number of words that they had read.

The number of words that they had read from the first advertisement was analysed by a between-subjects analysis of variance with the independent variables of typeface, type size, x-height, and reading skill. The use of a serif typeface led to better performance than the use of a sans serif typeface, but only for fast readers reading small typefaces and only for fast readers reading typefaces with a large x-height. The contrast between serif and sans serif typefaces had no significant effect for slow readers, for fast readers reading large typefaces, or for fast readers reading typefaces with a small x-height. These results suggest that any differences in the legibility of serif and sans serif typefaces will only arise as a result of very specific interactions with the effects of other features of the typeface and of the readers themselves.

In the studies mentioned in Sect. 5.4, Bartram (1982) and Rowe (1982) had only studied the connotations of typefaces when used for individual words. Even so, both maintained that these connotations would influence readers' interpretations when the typefaces were used for regular narrative. E. R. Brumberger (2003b) set out to test this idea by comparing the connotations of typefaces and the connotations of texts in which they were used. In her first study, she asked 80 students to rate how much each of 15 descriptors applied to each of 15 typefaces. A factor analysis of their responses yielded three broad dimensions, which she labelled "elegance," "directness," and "friendliness." Multidimensional scaling yielded a similar grouping of the 15 typefaces. Brumberger noted that the resulting categories were based on the semantic qualities of the typefaces, not their physical characteristics. In particular, each category subsumed both serif and sans serif typefaces (see also E. Brumberger, 2004; Mackiewicz & Moeller, 2004).

In her second study, Brumberger (2003b) presented another 80 students with 15 different texts, each containing 375 words, drawn from a variety of published sources. The participants were asked to read each text and to rate it on the same 15 scales. A factor analysis yielded three broad dimensions that were very different from those found in the first study. She labelled these new dimensions "professionalism", "violence", and "friendliness". Brumberger argued that she had demonstrated that readers consistently ascribe particular personality attributes to particular typefaces and text passages. However, the lack of concordance between the results of the two studies contradicts the idea that the connotations of typefaces affect readers' interpretation of the texts in which they appear.

Brumberger (2003a) selected one typeface that represented each of the dimensions in her first study: the cursive typeface CounselorScript for elegance, the serif typeface Times New Roman for directness, and the sans serif typeface Bauhaus Md

BT for friendliness. She also selected one text that represented each of the dimensions in her second study: a passage from a psychology textbook for professionalism, an excerpt from a novel for violence, and an excerpt from an article on snowboarding for friendliness. Different groups of students were presented with the texts in different typefaces in a between-subjects counterbalanced design and were asked to rate the relevant text on the 15 scales used in her earlier study. There were significant differences in the ratings given to the three texts, but no significant difference in the ratings given to texts in different typefaces. Brumberger suggested that the "persona" of the texts might have overridden that of the typefaces (p. 230).

Gasser et al. (2005) asked 149 psychology students to read an information sheet about tuberculosis that was being used at a local health-care facility. They were presented with the information sheet in one of four typefaces: a slab serif typeface with monospacing (Courier), a serif typeface with proportional spacing (Palatino), a sans serif typeface with monospacing (Monaco), or a sans serif typeface with proportional spacing (Helvetica). (With monospacing, each character occupies the same width, but with proportional spacing different characters take up different amounts of horizontal space.) They read the material silently at their own pace and were then given a short attitudinal questionnaire as a distractor task. Finally, they answered six open-ended questions on key points in the material. The students who read the material in serif typefaces answered more of the questions correctly than did those who read it in sans serif typefaces, although the difference only just attained statistical significance. Gasser et al. suggested that the students had found information printed in serif typefaces easier to read (and thus easier to remember) because they were more familiar with such typefaces in their regular educational material.

Juni and Gross (2008) asked 102 university students to read two satirical articles from the *New York Times*; one was concerned with government issues and the other with education policy. They then rated each article on 12 qualities. One of the articles was presented in the serif typeface Times New Roman, and the other in the sans serif typeface Arial. For the government article, the version presented in Times New Roman was rated as significantly more angry and as significantly less cheerful than the version presented in Arial. For the education article, the version presented in Times New Roman was rated as significantly more frivolous than the version presented in Arial. It should however be noted that Juni and Gross found just three significant differences out of a total of 24 comparisons between the two typefaces without controlling for the possibility of spurious results due to chance variation (i.e., Type I errors), suggesting that the choice of typeface generally made little difference to their participants' perceptions of the two articles.

6.5 Conclusions

It has been argued that the context of reading is a primary determinant of the legibility of different typefaces and the readers' expectations of the legibility of what they are reading. Newspaper headlines have been used as a specific context in which researchers have studied the legibility and connotations of different kinds of text. Wheildon (1990, 2005) presented an extensive programme of research on the legibility of different kinds of text. However, his research has come under extensive criticism and suffers from further issues that have not been noted in previous research. Several researchers have subsequently considered the effect of variations in typefaces and the expectations of readers in different kinds of situations. In general, research on reading in context provides no convincing evidence for any difference in the legibility of serif and sans serif typefaces. Nevertheless, there is a suggestion that readers' preferences and the connotations of serif and sans serif typefaces might well vary between different contexts.

Open Access This chapter is licensed under the terms of the Creative Commons Attribution 4.0 International License (http://creativecommons.org/licenses/by/4.0/), which permits use, sharing, adaptation, distribution and reproduction in any medium or format, as long as you give appropriate credit to the original author(s) and the source, provide a link to the Creative Commons license and indicate if changes were made.

The images or other third party material in this chapter are included in the chapter's Creative Commons license, unless indicated otherwise in a credit line to the material. If material is not included in the chapter's Creative Commons license and your intended use is not permitted by statutory regulation or exceeds the permitted use, you will need to obtain permission directly from the copyright holder.

Chapter 7
Younger and Older Readers

7.1 Younger Readers

Research with young children is of interest in the present connection. First, as novice readers, young children may be disproportionately sensitive to variations in typographic variables such as the presence or absence of serifs. Second, the preparation of educational material in either serif typefaces or sans serif typefaces might affect the facility with which children acquire the ability to read using such material.

Children's books were generally printed in serif typefaces until the 1930s (Walker & Reynolds, 2003). A notable exception was the educational material published by Nellie (Ellen) Dale in collaboration with the artist Walter Crane in the United Kingdom in the 1890s and early 1900s (Dale, 1902b, 1903; for discussion, see Brockington, 2012). Young children were introduced to individual letters and combinations of letters printed in a sans serif typeface and carried out various exercises that involved writing on blackboards or slates using coloured chalk as well as reading letters aloud (Dale, 1903, pp. 15–18). They were next introduced to a series of readers or "primers" containing groups of new words of increasing phonological complexity that were once again presented in a sans serif typeface. Each group of new words was then used in a short narrative that was printed in a *serif* typeface and accompanied by a relevant illustration. Examples are shown in Fig. 7.1 (see also Dale, 1902a; Walker, 2013, pp. 88–89). Dale's work was mentioned by Burt (1959, p. 8; 1960) in reviews of the literature, but her work now seems to have been almost completely forgotten.

Subsequently, many educational publishers produced their own typefaces consisting of "infant characters" modified for novice readers. For instance, the loops in characters such as *a* and *g* are only partially circular in sans serif typefaces (see the right-hand panel of Fig. 1.1 in Sect. 1.2) but are often more completely circular when rendered as infant characters. The assumption appears to have been that children should learn the simpler shapes of letters printed in a sans serif typeface in developing their handwriting before learning more complex shapes printed in a serif typeface (Coghill, 1980; Walker, 2013, pp. 31–35 et passim; Watts & Nisbet, 1974,

© The Author(s) 2022
J. T. E. Richardson, *The Legibility of Serif and Sans Serif Typefaces*,
SpringerBriefs in Education, https://doi.org/10.1007/978-3-030-90984-0_7

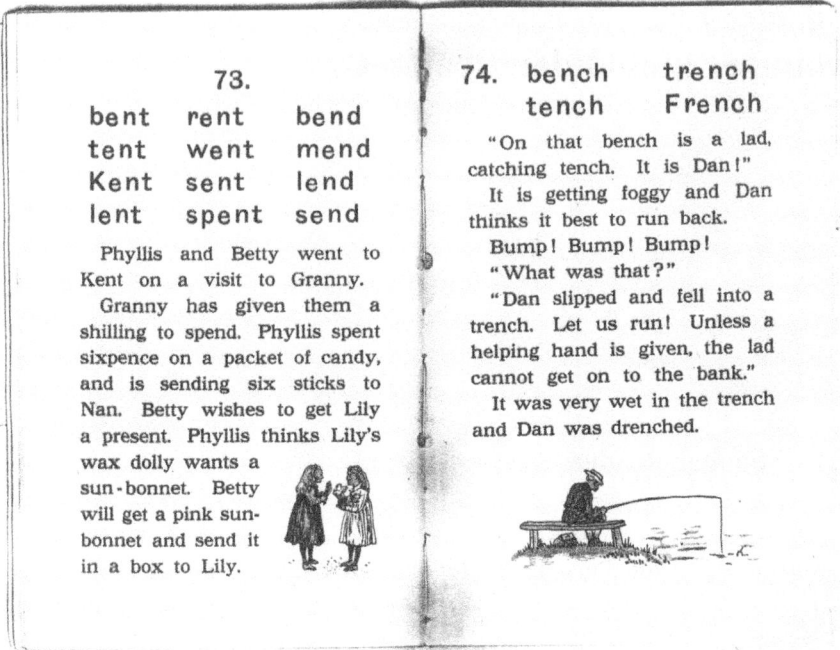

Fig. 7.1 Two pages from one of Dale's *Readers*. From *The Dale readers: Infant reader* (new ed.), by N. Dale. 1902. George Philip & Son. In the original, the illustrations and some letters in the sans serif headings were rendered in colour. Reproduced by kind permission of the Philip's Division of Octopus Publishing Group and Sue Walker from http://www.bookdata.kidstype.org/database/database/getImage?id=1018

p. 33). (This is, of course, exactly the pedagogical approach that had been promoted by Dale.) As a result, teachers nowadays tend to prefer sans serif typefaces over serif typefaces (Raban, 1984; Walker & Reynolds, 2003), and many books for younger readers are printed in sans serif typefaces (Bluhm, 1991; Walker, 2013, pp. 33–35).

7.2 Burt and Kerr's Research

One study often cited in support of the idea that serif typefaces are more legible than sans serif typefaces (e.g., Gallagher & Jacobson, 1993; Schriver, 1997, p. 274) was described by a psychologist, Cyril Burt, who collaborated with a physician, James Kerr. The research involved ten groups of children, each consisting of 15 boys and 15 girls aged 10–11 years. It was apparently carried out between 1913 and 1917, but the findings were not reported until after Burt's retirement in the 1950s (Burt, 1959; Burt et al., 1955). The final experiments used ten different serif typefaces. Burt et al. (1955) mentioned that they had carried out initial experiments with four other typefaces, including a sans serif typeface, but that these had proved "much inferior" to those that had been selected (p. 32). Burt (1959) explained that "in our own early experiments Dr Kerr and I [7] found almost at once that, for word recognition, a sans serif type face was the worst of all" (p. 9).

The reference in square brackets was to Kerr's (1926) account of his own findings. Kerr mentioned the possibility of using letters in a sans serif typeface, but he simply asserted, without argument or evidence, that "owing to irradiation they are not as legible as letters with thicker ends" (p. 552). (On the topic of irradiation, see Sect. 1.2.) No further information was provided regarding experimental comparisons between serif and sans serif typefaces in either Burt's account or Kerr's. Moreover, Hearnshaw (1979, pp. 227–261) concluded that there were serious doubts about the validity and authenticity of the data presented in Burt's publications, while Hartley and Rooum (1983) argued that this was true in particular of his work on typography. In fact, Burt et al. (1955, p. 32) claimed that the sans serif typeface which they had used was Gill Sans (see also Burt, 1959, p. 9), but this was not made available until 1928 (Kinross, 1992, pp. 62–63), which was more than a decade *after* Burt and Kerr had supposedly carried out their research.

7.3 Zachrisson's Research

The first systematic analysis of the legibility of serif and sans serif typefaces among children was carried out by Zachrisson (1965, pp. 97–108). Groups of 36 boys in Grade 1 (aged 7–8 years) were drawn from two Swedish schools. At one school, reading instruction was based on material printed in a serif typeface; at the other school, it was based on material printed in a sans serif typeface. The boys were tested individually and were asked to read aloud two pages of text. They were randomly assigned to receive text that was printed in one of two serif typefaces (Bembo or Nordisk Antikva) or one of two sans serif typefaces (Gill Sans or Mager Konsul). Zachrisson analysed the number of errors made when reading the text. There was a significant variation among the four typefaces, but the overall difference between the serif typefaces and the sans serif typefaces was not significant, and the difference between the two schools (and hence the typefaces used for reading instruction) was not significant.

In a subsequent experiment (pp. 109–115), Zachrisson drew a sample of 24 boys and 24 girls from Grade 4 (aged 10–11 years) of the school where instructional material was printed in a sans serif typeface. They were asked to read silently two different passages, but their reading was interrupted from time to time by comprehension questions. Once again, the passages were printed in one of two serif typefaces (Bembo or Fairfield) or one of two sans serif typefaces (Fin Grotesk or Gill Sans). Each participant read one passage in a serif typeface and the other in a sans serif typeface. There was no overall difference in the time taken to read each passage between the serif typefaces and the sans serif typefaces, and there was no significant variation among the four typefaces.

For his next experiment (pp. 115–121), Zachrisson presented individual words in isolation by means of a tachistoscope; alternate words were presented in a serif typeface (Mediaeval) and in a sans serif typeface (Mager Futura). The sample consisted of 12 boys in Grade 1 at each of the schools involved in the first experiment, together with six boys and six girls in Grade 4 at the school involved in the second experiment. The words were presented for 40 ms to the younger children and for 20 ms to the older children. Zachrisson analysed the number of words correctly reported, with partial credit for words that were incorrectly reported. There was no significant difference between the two typefaces for either the younger children or the older children, and no significant difference between the two schools.

Zachrisson employed the same research design and materials in a further study that employed Weiss's (1917) focal variator (pp. 121–124), which was described in Sect. 1.2. In this case, he analysed the threshold at which the blurred image of a word was correctly recognised. The difference between the two typefaces was not significant for either the younger children or the older children. The difference between the two schools for the children in Grade 1 was highly significant, but Zachrisson did not report the direction of the difference. Zachrisson also adopted the same research design using a perimeter, which is a device for presenting visual material in peripheral vision (pp. 124–128). This study failed to yield any significant results, and he inferred that this was not a useful way to measure the legibility of typefaces.

Zachrisson had also asked the children who participated in his first two experiments to rank order the four typefaces in which the passages had been presented (pp. 131–132). They were instructed as follows: "The point is to say which of these you find most legible—inviting, pleasant, easy, to read, and in which order you want to put them according to their legibility. Which one do you like best, next best, third and least?" (p. 131). There was no significant difference in the ranks assigned to the four typefaces, no significant difference between the children in Grade 1 and the children in Grade 4, and no significant difference between children in Grade 1 at the two schools (and hence between the typefaces used for reading instruction). Zachrisson concluded on the basis of all his findings that "*there is no significant difference, in objectively measured legibility, or subjective opinion regarding ease of reading between the OF* [serif] *and SS* [sans serif] *type faces*" (p. 132, italics in original).

7.4 Other Research with Children

Weiss (1978, 1982) asked 145 boys and girls in Grades 3 and 6 (aged 8–9 and 11–12) at two public schools in New York City to express their preferences among printed material that differed in page size, type size, and typeface. Each example was printed as a two-page spread of text but consisted simply of familiar words in a random sequence together with an arbitrary illustration presented in the top, middle, or bottom of the right-hand page. Weiss chose three typefaces that were easily discriminable: a sans serif typeface (Futura) and two serif typefaces (Paladium and Parinesy). The children were divided into three ability groups based on their scores in an achievement test and were interviewed individually about their perceptions and preferences regarding the different page sizes, the different typefaces, and the different positions of the illustrations.

Of the three factors, the typeface was regarded as important by children in Grade 3 but not by children in Grade 6, by boys but not by girls, and by children of medium ability but not by children of low or high ability. Children of low and middle ability tended to prefer the Futura sans serif typeface, whereas children of high ability tended to prefer the Paladium serif typeface. There was no significant difference in preferences between boys and girls or between the children in different grades. When asked to give reasons for their preferences, the children mainly referred to the legibility and the attractiveness of the printed material. As Weiss noted, this pattern is consistent with the results obtained by Tinker and Paterson (1942) in the case of adult readers.

Coghill (1980) carried out an informal study in which 38 children aged about 5 years who were being taught to read using materials printed in a sans serif typeface (Gill Sans) were asked to read aloud sentences printed in that typeface or in a number of serif typefaces: "In almost every case the children found little difficulty in reading alternative typefaces" (p. 257), and any reading errors tended to be repeated across the different typefaces.

Sassoon (1993) described a study in which 50 8-year-old children were shown a short passage in four different typefaces and four different settings and were asked to choose the typeface that they liked best and found easiest to read. There were two serif typefaces, Times Roman and Times Italic, and two sans serif typefaces, Helvetica and a slanting sans serif typeface that Sassoon herself had developed. Their preferences were fairly evenly distributed, and Sassoon argued that children were able to assimilate different typefaces as a result of their exposure to television graphics and other kinds of advertising. These findings led her to promote her own typeface, Sassoon Primary, in material for young readers (Bluhm, 1991). However, using two different procedures, Wilkins et al. (2009) found that children read words in this typeface less quickly (both aloud and silently) than they read words of the same x-height in the sans serif typeface Verdana. Wilkins et al. suggested that this was because, with Sassoon, neighbouring letters tended to use strokes that were more similar in shape.

De Lange et al. (1993) asked 160 schoolchildren to read two pages of text and to mark all occurrences of a particular word. Half received two pages in the same serif typeface (Times Roman), but the other half received the first page in Times Roman and the second page in a sans serif typeface (Helvetica). Each group contained equal numbers of children from each of four schools, equal numbers of children from Years 4 and 6, and equal numbers of children with high and low academic performance. De Lange et al. calculated the scanning speed on each page by dividing the number of marked targets by the scanning time and then calculated the gain in scanning speed from the first page to the second. There was no sign of any significant difference between the two conditions in the gain scores obtained by the children in either year. De Lange et al. concluded that there was no significant difference between the legibility of Times Roman and Helvetica as measured by a scanning process.

Walker and Reynolds (2003) presented excerpts from a children's reading book to 24 children aged 5–7 years in either a serif typeface (Century) or a sans serif typeface (Gill Sans) with or without infant characters. The typefaces were balanced across the four excerpts in different children. Walker and Reynolds measured the time taken to read each excerpt aloud as well as the errors that the children made in doing so. There was no significant variation in the time taken or in the number of errors made across the four typefaces. Walker and Reynolds took these results to confirm Coghill's (1980) view that children do not find non-infant characters problematic. The children were also asked to express their preferences among the typefaces: eight expressed no preference, but eight expressed a preference for Gill Sans without infant characters, and five expressed a preference for Gill Sans with infant characters.

Ripoli (2015) noted that in many Spanish-speaking countries children are taught to read using material printed in a cursive typeface before moving on to serif and sans serif typefaces. He tested 115 children who had been taught to read using a cursive typeface in the final year of preschool. They were asked to read aloud six short texts taken from a children's book in Spanish, each in one of six typefaces: a cursive typeface (Escolar 1), two serif typefaces (Sylfaen and Times New Roman), and three sans serif typefaces (Arial, Lexia Readable, and Comic Sans). Assignment of the six typefaces to the six texts was counterbalanced across different children, and the x-height of the lowercase letters in each typeface was matched to that of Arial 14-point type. There was significant variation across the six typefaces in the number of incorrect line breaks made by the children, but no significant variation in the number of words read correctly per minute or in the number of errors that they made. Ripoli observed that, despite having been taught to read using a cursive typeface, the children had no difficulty reading using non-cursive typefaces, even using Lexia Readable (which they had not previously encountered). There was also no difference in their performance on the serif typefaces and on the sans serif typefaces.

Griffiths (2020) devised the Comparative Rate of Reading Speed Test. This involved two displays, each consisting of 13 lines with 60 characters in each line. The characters were random groups of between one and seven lowercase letters. The first display was printed in black in the serif typeface Times, and the second display was printed in teal in the sans serif typeface Gill Sans. A total of 92 children aged 11–12 years were asked to read the characters aloud and were timed on their reading of the fifth line in each display. The mean times were 40.53 s for the Times display and 34.81 s for the Gill Sans display. The difference between these mean times was highly significant. Griffiths commented that the use of teal for the Gill Sans display had been "a concession to light sensitive subjects" (p. 11). He suggested that the result was due to binocular deficiency (i.e., unstable co-ordination of the eyes), but this only affects 15% of the general population (Hargreaves, 2008). He acknowledged that the effect of typeface had been confounded with that of colour, and he argued that the latter was the more important variable, possibly due to a reduction in contrast. However, since all of the children saw the displays in the same order, the result might simply have represented a practice effect.

7.5 Letter Reversals

It has been known for more than a century that children who are learning to read tend to confuse pairs of lowercase letters that are mirror images of each other (e.g., *b* and *d*; *p* and *q*) (Mach, 1897, p. 50). This appears to be true in all cultures that use the Western alphabet in reading and writing (Goikoetxea, 2006). It is mainly apparent in the final year of kindergarten and the early years of compulsory education, and it is apparent in a variety of tasks extending beyond reading aloud and writing to dictation (Thompson, 2009). Such errors are found in normal readers as well as in children with learning disabilities and children who are dyslexic; nevertheless, they become less common in older children as their reading develops (Cossu et al., 1995; Davidson, 1936; Kennedy, 1954). (They are sometimes seen in older neurological patients and occasionally in healthy older adults: Balfour et al., 2007.)

In the regular forms of most sans serif typefaces, the relevant letter pairs are exact mirror images of one another (see Fig. 1.1 in Sect. 1.2). Some authors have argued that the addition of serifs enhances legibility by making individual letters more discriminable from one another (e.g., Legros, 1922, p. 11; McLean, 1980, pp. 42–44). Yule (1988) and Wiebelt (2004) argued in particular that serifs help to differentiate between confusable letter pairs because they are no longer exact mirror images of each other. (For instance, the left-hand panel of Fig. 1.1 shows that, in both the letters *b* and *d*, the serifs are on the left-hand side of the ascenders.) However, other authors have argued that, even with the addition of serifs, the relevant pairs of lowercase letters do not differ appreciably from each other (e.g., Potter, 1949, p. 11). Indeed, Lockhead and Crist (1980) found that more explicit cues were needed to enable children to discriminate between such letter pairs.

Evaluating the two positions is difficult, because most researchers who have described letter reversals in young children have not specified the typeface used in their reading tests. It is therefore impossible to say whether the children had been asked to read letters printed in serif or sans serif typefaces. Popp (1964) did specify the use of a serif typeface (Century) to present lowercase letters in a two-alternative forced-choice experiment where children had to match letters presented on a touch-sensitive projection screen. Their error rate was highest on the pairs *b*–*d* and *p*–*q*, thus confirming that mirror reversals still occur with letters that are presented in a serif typeface.

In the study mentioned in Sect. 7.4, Ripoli (2015) examined the number of letter reversals made by 115 children when reading six texts printed in six different typefaces. The number of letter reversals was least for texts printed in the cursive typeface Escobar 1 and the sans serif typeface Lexia Readable, where there are additional cues that serve to differentiate the critical pairs of lowercase letters. However, there was no significant variation in the number of letter reversals for texts printed in the other four typefaces: two serif typefaces (Sylfaen and Times New Roman) and two sans serif typefaces (Arial and Comic Sans). These results imply that, in the absence of additional cues, the presence of serifs does not in itself help children to discriminate between pairs of letters that are otherwise mirror images of each other. In general, mirror reversals in children's reading and writing relate to structural properties of the relevant alphabet and not to the particular typeface used to render that alphabet (Treiman et al., 2014).

7.6 Older Readers

Vanderplas and Vanderplas (1980) suggested on the basis of interviews carried out with older people that many did not read as much as they would have liked because of difficulties with illegible type, although some publishers do produce large-type versions of newspapers and books intended for older readers. The researchers asked 28 volunteers aged between 60 and 83 to read 30 passages of 30–33 lines taken from Samuel Butler's novel, *Erewhon*. The 30 passages were presented in one of five type sizes from 10 to 18 points and in one of six typefaces. Three were serif typefaces (Century Schoolbook, Times Roman, and Bodoni), and three were sans serif typefaces (Helvetica, Spartan, and Trade Gothic). The type sizes and the typefaces were counterbalanced, but the order of the passages reflected the narrative structure of the novel. After reading each passage, the participants were given a short test of their comprehension to ensure that they had actually read the material, and they then rated the passage on six aspects of its presentation using a 7-point scale.

Their reading speed was significantly faster for passages with serif typefaces than for passages with sans serif typefaces, but there was also significant variation in reading speed across the six typefaces: Century Schoolbook yielded the fastest reading speed, but Spartan yielded the slowest. Their reading speed generally tended to increase with the type size. The participants also rated passages with serif typefaces more positively than passages with sans serif typefaces on their apparent size, how easily they could be read, and how easily they could be understood. They also rated 12-point typefaces as being the easiest to read.

One situation in which older readers encounter problems is in reading labels on their medication, regardless of whether the labels are prepared using dot-matrix printers (Zuccollo & Liddell, 1985) or more advanced laser printers (Watanabe et al., 1994). Smither and Braun (1994) asked 19 younger adults (mean age 25.48 years) and 20 older adults (mean age 71.05 years) to read the labels on 18 prescription bottles printed in a serif typeface with proportional spacing (Century Schoolbook), a monospaced slab serif typeface (Courier), or a sans serif typeface with proportional spacing (Helvetica). They read more slowly and made more errors on the labels printed in Courier than on the labels printed in either Century Schoolbook or Helvetica. The older adults also read more slowly and made more errors than the young adults when reading labels printed in Courier.

Smither and Braun suggested that the participants might have had problems because of the curvature of medication bottles. They repeated their experiment with new participants and with the 18 labels placed on a flat surface. These participants once again read the labels printed in Courier more slowly, but they did not make more errors than on the labels printed in Century Schoolbook or Helvetica. The older adults read more slowly than the young adults across the board, but they did not make more errors. Smither and Braun inferred that reading medication labels was more effective if they were printed with proportional spacing (Century Schoolbook or Helvetica) than if they were printed with monospacing (Courier). However, the presence or absence of serifs seemed to have little or no effect on the legibility of medication labels.

7.7 Conclusions

As novice readers, young children may be disproportionately affected by different typefaces. The use of different typefaces may also affect how readily children acquire the ability to read. Research by Burt (1959), Burt et al. (1955) and Kerr (1926) is often cited in support of the idea that serif typefaces are more legible. However, their accounts are inadequate and contain many contradictions. Zachrisson (1965) provided a more thorough account of the role of typographic variables in reading among children of different ages using various research methods and found no evidence for any difference in legibility between serif and sans serif typefaces. Subsequent research by other investigators has tended to confirm Zachrisson's conclusions. It has been known for more than 100 years that children tend to confuse letters that are mirror images of each other (such as p and q). This phenomenon occurs with both sans serif letters (which are true mirror images) and serif letters (which are not). Older readers tend to suffer from visual problems which may depend on typographical factors. This is of practical importance, as in the design of labels for medication containers. Nevertheless, there are no differences in the reading capability of older readers when presented with material printed in serif and sans serif typefaces.

Open Access This chapter is licensed under the terms of the Creative Commons Attribution 4.0 International License (http://creativecommons.org/licenses/by/4.0/), which permits use, sharing, adaptation, distribution and reproduction in any medium or format, as long as you give appropriate credit to the original author(s) and the source, provide a link to the Creative Commons license and indicate if changes were made.

The images or other third party material in this chapter are included in the chapter's Creative Commons license, unless indicated otherwise in a credit line to the material. If material is not included in the chapter's Creative Commons license and your intended use is not permitted by statutory regulation or exceeds the permitted use, you will need to obtain permission directly from the copyright holder.

Chapter 8
Readers with Disabilities

8.1 Readers with Visual Impairment

Nolan (1959) tested 264 children with visual impairment aged between 8 and 20. Half of the children counted as legally blind, but the others did not. Within each group, the children were randomly assigned to different subgroups to read material presented in either 18-point type or 24-point type in either a serif typeface (Antique with Old Style) or a sans serif typeface (Metrolite Medium). The material itself consisted of 72 paragraphs, each of about 90 words; after reading each paragraph, the children had to answer a short comprehension question by choosing one of five alternative words. Nolan measured the number of paragraphs that they had read in 30 min, statistically adjusted for variations in their reading comprehension.

The results showed that the children with better vision read faster than the children with poorer vision, and that the material presented in Antique with Old Style was read more quickly than the material presented in Metrolite Medium There was no significant difference in reading speed between the two type sizes, and there were no significant interactions among the effects of reading ability, type size, and typeface. Nolan commented that Metrolite Medium had needed a greater line length than Antique with Old Style, and she suggested that it was this feature rather than the absence of serifs that explained the slower reading time. In fact, Nolan had standardised the line length at 8.5 in. (21.6 cm) in all four sets of material. Consequently, the material that was presented in Metrolite Medium would have needed more lines, which suggests that it was the number of lines rather than the presence or absence of serifs that had given rise to differences in reading performance between the two typefaces.

8.2 Shaw's Research

During the 1960s, Alison Shaw was commissioned by the UK Library Association (now the Chartered Institute of Library and Information Professionals) to study the reading capabilities of people designated as *partially sighted*, defined as those who had difficulty reading normal sized book print but whose sight could not be fully corrected by the use of spectacles. The findings of the research were published as a formal report (Shaw, 1969). Shaw also published three short journal articles about the study, but these omit key information about her research methods and data analyses, and so this account is based on her formal report.

Shaw decided to investigate her participants' capacity for reading continuous printed text at a close reading distance (p. 9). She chose to investigate separate samples of adults and children, with a primary focus on the former (p. 17). She identified four aspects of the text which might be important: typeface, type weight, type size, and type spacing. With regard to typeface, she compared Plantin, a serif typeface which had been developed in 1913 by the Monotype Corporation, and Gill Sans, the sans serif typeface developed in 1928 by Eric Gill. She argued that these were representative examples of serif and sans serif typefaces that were relatively similar in terms of their x-height and their width (p. 23). With regard to weight, she compared the medium and bold versions of these two typefaces. With regard to size, she used 12-, 14-, 16-, 18-, 20-, and 24-point sizes. Each participant was tested to determine their visual acuity, and they received material in two successive point sizes that were just below and just above their acuity threshold (e.g., 16-point and 18-point sizes) (p. 24). With regard to spacing, she used four combinations of inter-letter, inter-word, and inter-line spacing (pp. 24–25), which yielded 32 different combinations of typeface, weight, size, and spacing.

Each participant received four of these 32 combinations. The materials consisted of short passages of common words that were combined to yield grammatical but semantically anomalous sentences (e.g., "Hungry bridges describe expensive farmers": p. 32). Each of the passages consisted of six sentences printed in five lines containing 38–41 characters. One passage was assigned to each of the 32 conditions, but a pre-test was carried out using ten readers with normal vision to ensure that the 32 passages were of comparable difficulty (p. 33). The conditions assigned to different participants and the order of their administration were counterbalanced using Graeco-Latin squares (p. 29).

Shaw measured the number of words in each passage that were read aloud correctly and the time taken to read them. She used this information to calculate the average time per correct word and normalised the latter by dividing it by the average time per word taken by the normal readers to read each passage. There was a positive correlation between the mean and the variability of the normalised data, and so the logarithms of these data were used. Finally, the four transformed values for each participant were expressed as deviations about their mean value in order to eliminate differences among the participants (p. 37). Multiple regression analyses

8.2 Shaw's Research

were then used to identify the factors that were responsible for significant variations in performance.

A total of 288 adults with visual impairment were recruited through government agencies and voluntary associations. Shaw compared their characteristics with those of the population according to national statistics, and she concluded that the sample was broadly representative of the national population of adults with partial sight (p. 53). The multiple regression analysis found that an increase in type size from the smaller size to the larger size led to an improvement of reading performance of 16%; an increase in the type weight from medium to bold led to an improvement of 9%; changing the typeface from serif to sans serif led to an improvement of 4%; and variations in spacing did not lead to a significant change in reading performance (p. 50). Shaw noted that her findings contradicted "traditionalist views that a serif face is always more legible than a sans serif" (p. 61), and she concluded that the effect of differences between the two typefaces was of minor importance compared with the effects of size and weight (p. 65).

Shaw compared various subgroups in the sample of adults with visual impairment. The most common causes of visual impairment (some adults had more than one cause) were:

- **Cataract**. This condition involves progressive cloudiness in the lens of the eye.
- **Glaucoma**. This condition involves damage to the retina or optic nerve.
- **Macular degeneration** (often known as age-related macular degeneration). This causes impaired vision in the centre of the visual field.
- **Myopia** (short-sightedness). In its severe form, this may be due to glaucoma or retinal detachments.

Comparing these subgroups, changing the typeface from serif to sans serif had significantly benefited participants with macular degeneration but not the other three subgroups (p. 50). Independent of this, 72 exhibited a congenital impairment, having been affected since birth or early childhood, whereas 112 exhibited an acquired impairment, having been affected since the age of 50 (p. 43). Changing the typeface from serif to sans serif had benefited the latter but not the former.

Shaw also recruited 48 children with visual impairment who had a reading age of at least 11 from schools for partially sighted children. Once again, Shaw compared their characteristics with those of the national population, and she concluded that they were representative of older, brighter children in schools for partially sighted children (p. 56). The multiple regression analysis found that an increase in type weight from medium to bold led to an improvement of reading performance of 7%, but that variations in type size, typeface, or spacing did not lead to a significant change in reading performance (p. 50). The most common causes of visual impairment were myopia, cataract, and nystagmus (which causes visual impairment by disrupting the normal pattern of eye movements), but these subgroups were too small for any further statistical analysis. After they had read the passages, the children were asked which of the four examples they thought was the clearest. There was no correlation between the children's personal subjective judgement and their objective reading performance (p. 49).

Shaw concluded that the sans serif typeface was slightly more legible than the serif typeface for her adult readers, but that there was no measurable difference for the children (p. 65). This might be taken to suggest that the advantage of the sans serif typeface was due to the adults' increased familiarity and experience in reading such typefaces. However, there are two problems with Shaw's study. First, although she carried out a large number of statistical tests on her data, she did not control for the possibility of Type I errors. As a result, some of the statistically significant differences which she found might have been spurious results due to chance variation.

Second, and more fundamentally, Shaw's focus on readers with visual impairment meant that she ignored the performance of the normal readers in her study. In normalising the performance of her participants on each passage against the performance of normal readers on the same passage, she ignored the possibility that the normal readers themselves might have shown significant differences between the serif and sans serif typefaces. Suppose that the readers with visual impairment produced reading speeds of 50 words/min on a passage printed in Plantin and 60 words/min on a passage printed in Gill Sans, and that the normal readers produced reading speeds of 100 words/min and 120 words/min on these passages. Normalising the former data would have yielded scores of 0.5 for both typefaces. (Taking logarithms and deviations about the mean value would not have changed this situation.) In other words, the normalisation process might well have obscured differences in legibility in *both* normal readers *and* those with visual impairment. Unfortunately, Shaw did not present the descriptive statistics to enable readers to establish whether or not this was a possibility.

8.3 Children in Special Education

Pittman (1976) compared 48 children with learning disabilities and 48 nondisabled children in their reading comprehension. She showed them stories consisting of five paragraphs and then administered ten sentence-completion questions. Each story was presented over four trials. The children in each group received stories in one of four typefaces: one was in a serif typeface (Pica), two were in sans serif typefaces (Gothic and Primary), and one was in a cursive typeface (Script). Not surprisingly, the children's performance increased over the four trials. The nondisabled children obtained higher scores than did the children with learning disabilities. However, there was no significant difference among the children' scores on the four typefaces. Pittman concluded "that the style of type is not an important variable in the reading comprehension of LD [learning-disabled] and normal children" (p. 115). It might be noted that the results obtained by the comparison group of nondisabled children confirm the conclusion of Chap. 7 that there is no difference in the legibility of serif and sans serif typefaces among normal young readers.

Section 7.4 mentioned a study by Sassoon (1993) in which nondisabled children were shown a short passage in four different typefaces and were asked to choose the typeface that they liked best. Sassoon repeated this study with 50 children who were in special education and aged between 8 and 13. Apart from describing them as having "learning difficulties" (p. 158), she did not mention what their special needs actually were. In this case, there were clear differences: 44% chose the slanting sans serif typeface, 28% chose the serif Times Italic, 18% chose the sans serif Helvetica, and 10% chose the serif Times Roman. Sassoon attributed the low preference for Times Roman to the fact that its pronounced serifs and short descenders affected the identification of certain letters. However, in Sect. 7.4 it was noted that Sassoon had developed the slanting sans serif typeface herself and had been promoting it for use in material for young readers. It would have been better if Sassoon had employed assistants who were blind as to the specific research hypotheses to avoid the possibility of researcher bias.

Haugen (2010) tested 14 children in special education in Grades 3–6 (aged 8–11). Once again, apart from describing them as having "mild special education needs" (p. 17), she did not discuss what their special needs actually were. She asked them to read aloud four passages of between 260 and 440 words printed in different typefaces: Bookman, a serif typeface; Comic Sans, a sans serif typeface based on comic-book lettering; Helvetica, a more regular sans serif typeface; and Times, another serif typeface. Each passage took about 5–10 min to read. However, Haugen did not time the children exactly but instead monitored their behaviour. Finally, she showed them all four of the passages that they had read and asked them to say which style of letters they had found the easiest to read and which they liked the best (pp. 83–93).

All the children completed all four passages, but they appeared more restless when reading the passages in Comic Sans and Times than when reading the passages in Bookman and Helvetica. Haugen commented that the issue was not the difference between serif and sans serif typefaces but the design of each typeface when compared with that of the others. The number of words read incorrectly was similar across the four typefaces, although Times yielded the fewest while Comic Sans yielded the most. The children were more likely to skip words printed in the two serif typefaces than those printed in the two sans serif typefaces, whereas they were more likely to pause when reading words printed in the two sans serif typefaces than when reading words printed in the two serif typefaces. Finally, they were more likely to run together two successive sentences printed in Comic Sans than those printed in the other typefaces (pp. 121–128).

Nine of the 14 participants thought that one of the serif typefaces was the easiest to read, and five thought that one of the sans serif typefaces was the easiest to read, four of whom chose Comic Sans. However, their choice did not seem to bear any relationship to their actual reading performance. Moreover, 12 of the 14 participants chose a sans serif typeface as the one they liked the best (of whom eight chose Comic Sans), one chose a serif typeface, and one insisted on choosing a serif typeface *and* a sans serif typeface (pp. 129–139). Haugen argued that their preferences had been influenced by their prior experience of sans serif typefaces on game systems, computers, cellular phones, and other electronic devices (pp. 164–165).

8.4 Readers with Congenital Visual Impairment

Uysal and Düger (2012) evaluated the effects of a 3-month training programme in 35 children with visual impairment at a Turkish primary school. Their preferences for different typefaces were assessed before and after the programme by showing them a 20-word sentence in Turkish in five different typefaces (Arial, Comic Sans, Tahoma, Times New Roman, and Verdana) and ten different type sizes. Their reading speed was averaged across all the typefaces and sizes. Before the programme, they preferred the sans serif typefaces such as Verdana (11 children) or Arial (8 children) as opposed to the serif typeface Times New Roman (3 children). The programme initially adopted their preferred typeface but gradually incorporated the other typefaces. It led to a significant increase in both their reading speed and their writing speed, but not in the judged legibility of their handwriting. After the programme, most children indicated that they were comfortable with any of the typefaces except for the sans serif Tahoma.

Skilton et al. (2018) conducted a focus group that involved eight people with deaf-blindness. Such individuals are born with a hearing loss and develop visual impairment during their early childhood. The aim of the focus group was to identify the participants' accessibility needs for their involvement in future research. Their recommendations included the provision of printed materials in a large size (18-point or higher) and in a sans serif typeface such as Arial. However, Skilton et al. acknowledged that these were among the recommendations that were typically provided for improving the accessibility of information of deaf-blind people. Their account left it unclear whether these recommendations were based on their own negative experiences with serif typefaces or whether they were just repeating back conventional attitudes that they had previously acquired from figures in authority.

8.5 Readers with Acquired Visual Impairment

Prince (1967) suggested that the impact of typographical variables might be different in older people with acquired visual disorders. As mentioned in Sect. 8.2, Shaw (1969) had found an advantage for a sans serif typeface over a serif typeface for those with acquired visual impairment but not for those with congenital visual impairment.

Estey et al. (1990) showed 52 patients with an average age of 69.4 years admitted for cataract surgery a page of text printed in a 12-point sans serif typeface, Univers Medium. Only 65% of the patients said that they could read the text, while 35% found it blurry. Estey et al. then showed the patients two pages of text: One was printed in 14-point Univers Medium, and the other was printed in a 14-point serif typeface, Century Schoolbook. Only 2% of the patients said that they could not read these texts. Of the 52 patients, 65% said that they preferred Univers Medium, 33% said that they preferred Century Schoolbook, while 2% had no preference. Estey et al. argued that material for patients with visual deficits should be printed in 14-point sans serif typefaces.

8.5 Readers with Acquired Visual Impairment

Campbell et al. (2006) carried out two studies in people with age-related macular degeneration (AMD). Participants were recruited from the members of the Canadian National Institute for the Blind (now the CNIB Foundation). They were asked to compare samples of text printed in six different typefaces using reading aids if necessary. Two were serif typefaces (Times Roman and Lucida), while four were sans serif typefaces (Adsans, Arial, Clearview, and Verdana). Adsans had been devised in 1959 to be used in a small (4.75 point) size in newspaper classified advertisements. It is not available in common computer applications, but it was used as the basis for Verdana. Clearview was devised in the 2000s for use on road signs. In both studies, the participants rated how easy it was to read each typeface on a 7-point scale from "impossible to read" to "very easy to read." In the first study, they also ranked the samples from the easiest to the hardest to read. However, this proved to be rather demanding, and so ranking was not used in the second study.

In the first study, 241 participants aged 50 or older were shown excerpts from Robert Louis Stevenson's novel, *Treasure Island*, all printed in 16-point type. The passage printed in Adsans was given significantly higher ratings than any of the other samples, and it was ranked the easiest to read by more than 50% of the participants. In the second study, 157 participants were shown information leaflets for over-the-counter medicines amended to refer to unfamiliar products and printed in 7-point type. The leaflet that was printed in Adsans was once again given significantly higher ratings than any of the other samples. Campbell et al. remarked that in both studies Times Roman had been given low ratings despite being a relatively familiar typeface, which suggested that the participants were not simply rating the typefaces on the basis of their familiarity. These findings show that people with AMD have a preference for Adsans, but they do not constitute evidence with regard to its objective legibility.

Rubin et al. (2006) asked 43 patients with mild cataract or glaucoma to read texts printed in four different typefaces. One was Tiresias, a sans serif typeface developed for people with impaired vision by the Royal National Institute of Blind People in the United Kingdom. This was compared with the serif typeface, Times New Roman, and two other sans serif typefaces, Foundry Form Sans and Helvetica. Their reading speed was found to be significantly faster with Tiresias than with the other three typefaces. However, although nominally of the same point size, the four typefaces occupied different amounts of horizontal and vertical space. When this factor was statistically controlled, the advantage of Tiresias disappeared. Rubin et al. concluded that variations in typeface had little influence on the reading speed of people with mild to moderate sight problems.

Tarita-Nistor et al. (2013) tested 24 patients with AMD using reading charts printed in Times New Roman, Courier, Arial, and a version of Andale Mono. These required them to read individual sentences presented at progressively smaller sizes. Times New Roman is a serif typeface, and Courier is a slab serif typeface, whereas Arial and Andale Mono are both sans serif typefaces. Times New Roman and Arial are both proportionally spaced, whereas Courier and Andale Mono are both monospaced. Tarita-Nistor et al. measured three aspects of their participants' performance: their reading acuity, which was the smallest print size that could be read without significant errors; the maximum reading speed, which was the highest speed at which text could be read without regard to print size; and the critical print size, which was the smallest print size that could be read with maximum speed. There was no significant variation among the typefaces in either critical print size or maximum reading speed, but there was significant variation in reading acuity: surprisingly, and—contrary to the results of the study by Smither and Braun (1994) that was mentioned in Sect. 7.6—text printed in Courier yielded significantly better reading acuity than text printed in the other three typefaces, but text printed in Arial yielded significantly worse reading acuity than text printed in the other three typefaces.

Hedlich et al. (2018) administered reading charts containing sentences printed in the slab serif typeface Courier New or the sans serif typeface Arial to 16 patients with visual impairment before or after cataract surgery. They found no significant difference between the two typefaces in their reading acuity, in their critical print size, or in their maximum reading speed. One limitation of this study, apart from the small sample size, was that the reading chart printed in Arial was always presented before the reading chart printed in Courier New, so that the researchers had no control over the effects of fatigue or practice. They also asked the participants about their preference between the two typefaces: eight of the participants preferred Arial, four preferred Courier New, and four had no preference.

Nersveen et al. (2018) carried out a postal survey of adults with a wide variety of visual impairments. They identified ten typefaces for consideration. Seven were printed in regular font, including two serif typefaces (Scala and Times Roman) and five sans serif typefaces (Frutiger, Helvetica, Scala Sans, Tiresias, and Verdana). Three were printed in bold font, including one serif typeface (Scala Bold) and two sans serif typefaces (Scala Sans Bold and Tiresias Bold). The participants were a random sample of 5,000 members of the Norwegian Association of the Blind and Partially Sighted. Each typeface was presented in five different body sizes (8, 10, 12, 14, and 16 points, but scaled so that their x-heights matched those of Times Roman), and in ten variations in contrast (from black type on white background to white type on black background), yielding 500 conditions. Each was presented as three or more lines of text, together with a 4-point rating scale in which the response categories were "Easily readable", "Readable with some difficulty", "Difficult to read", and "Unreadable". This yielded a booklet consisting of 50 printed pages.

The participants were instructed to carry out the task only if they were partially sighted and able to read printed text (with a magnifying glass or with supplementary lights if necessary). Completed booklets were returned by 830 participants. Repeated-measures tests were employed to compare the ratings given to the serif typefaces and the sans serif typefaces. For typefaces at 12 and 14 points, the difference was not statistically significant. For those at 8, 10, and 16 points, sans serif typefaces received significantly higher ratings than did serif typefaces. Nevertheless, the differences were small in magnitude and only achieved significance because of the very large sample size. A similar pattern emerged when comparing the ratings given to the Scala typefaces and the Scala Sans typefaces (J. Nersveen, personal communication, June 22, 2020).

8.5 Readers with Acquired Visual Impairment

Although Nersveen et al. described their experiment as a study of the legibility of printed text, their data actually consisted of the participants' ratings of the subjective acceptability of the different typefaces rather than any measure of their objective legibility. The apparent preference for sans serif typefaces is thus consistent with the findings of Estey et al. (1990) and Campbell et al. (2006), although the effect was far less pronounced. One problem is the response rate of only 16.6%. This might be partly explained by the fact that the participants did not receive any personal reward for carrying out the task, although two respondents chosen at random were given a "prize" of a Digital Audio Broadcasting radio worth 1,000 Norwegian kroner. As a result, most of the participants may not have been willing to devote time and effort to a rather burdensome task. Whatever the cause, it does suggest that the study suffered from sampling bias, in that the respondents might not have been representative of the target population.

8.6 Readers with Aphasia

The term *aphasia* covers a wide variety of disorders of spoken language, but a majority of people with aphasia also exhibit impairment of reading (Brookshire et al., 2014). Wilson and Read (2016) tested nine participants who had been diagnosed with mild-to-moderate aphasia as the result of cerebrovascular accidents. They were given a standardised test of reading comprehension that consisted of 35 short paragraphs. In each case, the participants had to choose one of four alternative words or phrases to complete the final sentence. For each participant, the paragraphs were randomly assigned to one of seven conditions. One involved the presentation of the original paragraph in a serif typeface (Times New Roman), and the other six involved different manipulations. For two manipulations, the typeface was amended either to a sans serif typeface (Verdana) or to an ornate cursive typeface (Harrington). The participants achieved significantly higher scores with the sans serif typeface than with either of the other two typefaces. Wilson and Read did not speculate as to why patients with aphasia might find serif typefaces less legible. Some researchers have found that people with aphasia prefer material printed in a sans serif typeface (Rose et al., 2011), but other researchers have not (Haw, 2017, p. 129; Herbert et al., 2019).

8.7 Readers with Dyslexia

The term *dyslexia* refers to a specific disorder of reading that may result from a wide variety of causes (and may be either congenital or acquired). For many years, the British Dyslexia Association (2018) has recommended that documents printed for people with dyslexia should use sans serif typefaces, and this has been taken to encompass teaching materials for students (Shaw & Anderson, 2017). However, the Association did not cite any evidence to support this recommendation. In fact, relevant evidence has been obtained in attempting to evaluate Dyslexie, a typeface that was developed by Christian Boer in 2008 to try to facilitate reading among children and adults with dyslexia (http://www.dyslexiefont.com). It is a sans serif typeface characterised by a relatively large x-height and relatively wide vertical and horizontal spacing between the letters. It is available under licence for both Microsoft Windows and Apple computer systems.

Marinus et al. (2016) recruited 39 children who were undergoing remediation for low progress in reading. They were asked to read aloud four passages, each of 200 words. One passage was presented in a 14-point Dyslexie typeface. A second was presented in 16-point Arial, a sans serif typeface which matched the x-height of the letters in the first passage. A third passage was presented in 16-point Arial with an overall increase in spacing. The fourth passage was presented in 16-point Arial with increased spacing both between words and between letters within words to match the spacing used in the first passage. The order of the conditions and the assignment of the passages to the conditions was counterbalanced across different participants. The children were scored on the number of words that they had read correctly per minute in each of the four conditions. Marinus et al. found that their reading speed in the first condition was significantly faster than in either the second or third, but that it was not significantly different in the fourth condition.

Kuster et al. (2018) carried out two experiments to evaluate the Dyslexie typeface. In the first experiment, 170 children with dyslexia were asked to read aloud two passages at separate sessions. One passage was presented in 12-point Dyslexie; the other was presented in 13-point Arial, adjusted to match the vertical spacing of Dyslexie. The order of the two conditions was counterbalanced across the participants. The children read significantly more quickly and made fewer errors on the second passage than on the first. However, there was no significant difference in either reading time or the number of errors between the typefaces.

In their second experiment, Kuster et al. tested 102 children with dyslexia and 45 children without dyslexia. They were each asked to read aloud three lists of words of varying complexity at three separate sessions. At each session, one list was presented in Dyslexie, whereas the other two lists were presented in the serif typeface Times New Roman and the sans serif typeface Arial, in both cases adjusted to match the x-height and the vertical spacing of Dyslexie. The order of the three conditions and the assignment of word lists to conditions was counterbalanced, and the children were scored on the number of words that they read correctly in one minute. Not surprisingly, the children without dyslexia obtained higher scores than the children with dyslexia, and performance varied inversely with the complexity of the words. However, there was no significant difference in the performance of either the children with dyslexia or the children without dyslexia across the three typefaces.

Finally, Powell and Trice (2020) recruited 36 children with dyslexia. They were each asked to read aloud three stories; each story contained 200 words and was followed by three factual questions to test the children's comprehension. One story was presented in 12-point Dyslexie; the others were presented in 14-point Arial and 14-point Times New Roman, both adjusted to match the horizontal and vertical spacing of Dyslexie. The order of the three stories and the assignment of the stories to the three typefaces was counterbalanced across different children. There was no significant variation across the three typefaces in terms of the mean time that the children took to read the stories, no significant variation in terms of the number of errors that they made, and no significant variation in their comprehension scores.

All three of these studies indicated that the effectiveness of the Dyslexie typeface is due to its increased spacing and not to its different letter shapes. If other typefaces are adjusted to match the spacing used for the Dyslexie typeface, the reading performance of children with dyslexia does not differ across the different typefaces. However, both Kuster et al.'s (2018) second experiment and Powell and Trice's (2020) study show in addition that the reading performance of children with dyslexia does not differ between serif and sans serif typefaces if they are matched in terms of their spacing. This contradicts the recommendation made by the British Dyslexia Association (2018) that documents printed for people with dyslexia should use sans serif typefaces.

8.8 Conclusions

Any differences in the legibility of serif and sans serif typefaces might become more apparent in readers whose visual systems are challenged as the result of disablement. In fact, the modal finding is that there are no differences in the reading capability of readers with a variety of disabilities when they are presented with material printed in serif and sans serif typefaces. It might be thought that children with congenital visual impairment would be more sensitive to typographical factors, but in fact such children rapidly adapt to reading both serif and sans serif typefaces. It has been suggested that the effects of acquired visual impairment might be different from the effects of congenital visual impairment, but both groups appear to be equally proficient in reading serif and sans serif typefaces. A majority of people with aphasia exhibit impairment of reading. In this field, it is often taken for granted that people with aphasia will find sans serif typefaces more legible, but there is only one study with a very small sample of participants that supports this position. Certainly, there is now good evidence that the reading performance of children with dyslexia does not differ between serif and sans serif typefaces when they are matched in terms of their spacing.

Open Access This chapter is licensed under the terms of the Creative Commons Attribution 4.0 International License (http://creativecommons.org/licenses/by/4.0/), which permits use, sharing, adaptation, distribution and reproduction in any medium or format, as long as you give appropriate credit to the original author(s) and the source, provide a link to the Creative Commons license and indicate if changes were made.

The images or other third party material in this chapter are included in the chapter's Creative Commons license, unless indicated otherwise in a credit line to the material. If material is not included in the chapter's Creative Commons license and your intended use is not permitted by statutory regulation or exceeds the permitted use, you will need to obtain permission directly from the copyright holder.

Chapter 9
General Conclusions to Part I

9.1 Key Findings from Part I

As was mentioned in Sect. 1.4, Part I has reviewed diverse studies using diverse methods of data collection, and this precludes any formal meta-analysis to integrate the findings. One must instead focus on the most common finding—the *modal* finding—regarding the legibility of serif and sans serif typefaces: superiority of serif typefaces; superiority of sans serif typefaces; or no difference.

The research question is whether there are differences in the legibility of serif and sans serif typefaces when they are used to generate printed material. The modal finding from the research studies that have been reviewed is that there are not. This applies to four out of six experiments on reading letters and words (Sects. 4.1 and 4.3); the two Korean studies yielded contradictory results. It applies to five out of six experiments on reading sentences (Sect. 5.1; see also Sects. 6.1 and 6.2) and to all four experiments on the comprehension of text, whether using measures of speed or accuracy (Sect. 5.2). It also applies to all eight experiments on the reading capability of younger readers and to two of the three experiments on the reading capability of older readers (Chap. 7). Two studies have been cited in support of the supposed superiority of serif typefaces, but these can be discounted: one failed to report any empirical data on the issue (Burt, 1959; Burt et al., 1955), and the other suffered from irredeemable methodological problems (Wheildon, 1990, 2005).

It is unfortunate that there has been relatively little work on the legibility of serif and sans serif typefaces in readers with disabilities as opposed to their subjective preferences for different kinds of typeface (Chap. 8). Two studies found no difference in legibility between serif and sans serif typefaces (Pittman, 1976; Rubin et al., 2006). One found a superiority for serif typefaces among children with congenital visual impairment (Nolan, 1959), but this study seems to have suffered from methodological problems. A fourth study found a superiority for sans serif typefaces among patients with aphasia (Wilson & Read, 2016). It is generally assumed that sans serif typefaces are more appropriate for people with aphasia, and there is an urgent need for more research to evaluate this assumption. Finally, two studies have found that the reading performance of children with dyslexia does not differ between serif and sans serif typefaces when they are matched in terms of the spacing of the letters.

9.2 Preferences and Connotations

With regard to research studies concerned with readers' preferences between serif and sans serif typefaces and the connotations of the two kinds of typeface, the modal finding is that among adult readers there is no overall preference between serif and sans serif typefaces, nor any overall difference in the connotations of serif and sans serif typefaces (Sect. 5.4). Even so, there is a suggestion that readers' preferences and the connotations of serif and sans serif typefaces may vary between different contexts (see Schriver, 1997, pp. 289–303; Zachrisson, 1965, pp. 156–62). This has major implications for educational publishing and educational assessment:

- For authors, editors, and publishers of books in many fields, any such differences will be mainly of commercial relevance. However, authors and editors of academic articles (and of books, too, in the humanities) will want to be assured that their work is evaluated in terms of its content rather in terms of its typographical appearance. This provides a far more logical reason for requiring that manuscripts should be submitted for publication to academic journals and publishers in a standard typeface than simply asserting that one kind of typeface is more legible than another.
- The issue of fairness is especially relevant in the context of academic assessment. It is possible that teachers and other assessors will give more positive evaluations of students' assignments if the teachers and students share the same typographical preferences than if they differ in those preferences (although there seems to be no empirical evidence on this matter). It would be useful if teachers who are responsible for particular course units (and, ideally, for entire degree programmes) could agree on their typographical preferences and make these known to their students.

There is a need for research on whether reviewers' evaluations of academic manuscripts and teachers' evaluations of students' assignments are affected by their own preferences and expectations. Nevertheless, the available evidence suggests that these variations in readers' expectations and preferences depend on their prior experience and familiarity with different typefaces and not on any intrinsic properties of the typefaces themselves. Indeed, the results that were obtained by Uysal and Düger (2012) indicate that even readers who are visually impaired will find most typefaces relatively congenial after a gradual period of exposure.

9.3 Implications for Previous Assumptions

Where does this leave previous assumptions about the legibility of serif and sans serif typefaces? There is no support for traditional beliefs that serif typefaces are superior to sans serif typefaces and certainly no support for Morison's (1959) assertion that "the serif is essential to the reading of alphabetical composition" (p. xi). Regarding the American Psychological Association's (2010) insistence that a serif typeface "improves readability and reduces eye fatigue" (pp. 228–229), Perea (2013) remarked: "There are no well-founded theoretical reasons to use of [sic] a serif font over a sans serif font—beyond subjective preferences" (p. 16). To this one might add: and there is no convincing empirical support, either.

The assertion contained in *Merriam-Webster's Manual for Writers and Editors* (1998) that "studies of typeface legibility have tended to demonstrate that standard serif typefaces can be read somewhat more easily and quickly than standard sans-serif typefaces" (p. 330) is factually incorrect. Finally, the length of the reference list at the end of this book contradicts Kullmann's (2015) statement that there have only been "sporadic" studies on this issue (p. 1), and one can certainly dismiss his assertion that previous research has not led to any clear conclusion. On the contrary, based on the wealth of evidence that has accumulated over the last 140 years, the clear conclusion is that there is no difference in the legibility of serif typefaces and sans serif typefaces when they are used to produce printed material.

9.4 The American Psychological Association's Current Position

The guidelines in the sixth edition of the Association's *Publication Manual* followed those in previous editions. However, a seventh edition was published while this monograph was being written; the new guidelines have already been adopted by the American Educational Research Association and are likely to be adopted by other organisations in the future. This seventh edition takes a rather different approach (American Psychological Association, 2020):

> APA [American Psychological Association] Style papers should be written in a font that is accessible to all users. Historically, sans serif fonts have been preferred for online works and serif fonts for print works; however, modern screen resolutions can typically accommodate either type of font, and people who use assistive technologies can adjust font settings to their preferences. Thus, a variety of font choices are permitted in APA style....
>
> Use the same font throughout the text of the paper. Options include
>
> - a sans serif font such as 11-point Calibri, 11-point Arial, or 10-point Lucida Sans Unicode or
> - a serif font such as 12-point Times New Roman, 11-point Georgia, or normal (10-point) Computer Modern....

> We recommend these fonts because they are legible and widely available and because they include special characters such as math symbols and Greek letters. (p. 44)

An accompanying background paper confirms that the focus of the new guidelines is on the accessibility of typefaces for users with disabilities rather than on their legibility per se (Accessibility, 2020). The paper also refers to the Web Content and Accessibility Guidelines produced by the World Wide Web Consortium, suggesting that it is concerned with reading from screens rather than reading from paper, although this is not made explicit. With regard to the legibility of serif and sans serif typefaces, the paper makes the following statement:

> It is a common misconception that serif fonts (e.g., Times New Roman) should be avoided because they are hard to read and that sans serif fonts (e.g., Calibri or Arial) are preferred. Historically, sans serif fonts have been preferred for online works and serif fonts for print works; however, modern screen resolutions can typically accommodate either type of font, and people who use assistive technologies can adjust font settings to their preferences.
>
> Research supports the use of various fonts for different contexts. For example, there are studies that demonstrate how serif fonts are actually superior to sans serif in many long texts (Arditi & Cho, 2005; Tinker, 1963). And there are studies that support sans serif typefaces as superior for people living with certain disabilities (such as certain visual challenges and those who learn differently; Russell-Minda et al., 2007). ("Myth 1," paras. 1–2)

The choice of reference citations in this statement is rather odd. First, in discussing different styles of typeface, Tinker (1963, pp. 46–48) referred to the study by Paterson and Tinker (1932), who used a speed-of-reading test to measure the legibility of each of ten typefaces. Seven were serif typefaces that had been nominated by a large number of editors and publishers as being worthy of study, of which Scotch Roman was used as a benchmark. The results showed that "type faces in common use do not differ significantly" (Tinker, 1963, p. 48). The other three were chosen in order to be "radically different" (p. 46): Kabel Light, a sans serif typeface; American Typewriter, a slab serif typeface that imitated typewriting; and Cloister Black, an elaborate serif typeface. Both American Typewriter and Cloister Black were read significantly more slowly than Scotch Roman, but Kabel Light was not. Tinker concluded: "Type faces in common use are equally legible.... A serifless type, Kabel Light, is read as rapidly as ordinary type" (p. 64). In other words, Tinker (1963) did *not* show that serif typefaces were superior to sans serif typefaces. Indeed, in a subsequent annotated bibliography, Tinker (1966, p. 84) strengthened his conclusion in the light of research findings since the study by Paterson and Tinker (1932): "Typefaces in common use are equally legible. This includes the typefaces with serifs and those without serifs."

Second, in addition to the study by Arditi and Cho (2005) that was mentioned in Sect. 5.1 and involved the presentation of "scrambled" text, these researchers carried out an experiment where they asked just four participants to read aloud individual sentences. The sentences were presented one word at a time on a computer screen. Arditi and Cho found no difference in performance between sentences in a slab serif typeface and sentences in a sans serif typeface. It should be noted that they did not make use of "long texts". However, the main point is that, once again, Arditi and Cho did *not* show that serif typefaces were superior to sans serif typefaces.

Third, the review by Russell-Minda et al. (2007) on the legibility of typefaces for readers with visual impairment covered both research on reading from paper and research on reading from screens. They did indeed conclude: "Sans serif typefaces, such as Arial, Helvetica, Verdana, or Adsans, are more readable than is Times New Roman, for example" (p. 413). However, this was not supported by the evidence that they described: they cited eight studies, of which six had found no significant difference in legibility between serif and sans serif typefaces. In fact, their abstract stated, "Research has not produced consistent findings" (p. 402). Moreover, in the original report on which their published review was based, Russell-Minda et al. (2006) had arrived at a very different conclusion: "Based on results from existing studies, the effects of the presence or absence of serifs on text legibility seem to be inconclusive" (p. 23). In short, they definitely did *not* demonstrate that sans serif typefaces were superior for people living with certain disabilities.

In other words, each of these three reference citations is in error because it fails to support the statement to which it is attached. In the bibliographic research literature, these are referred to as "quotation errors", although they include indirect quotations, paraphrases, and summaries as well as direct quotations. Mertens and Baethge (2011) demonstrated that around 20% of reference citations in the medical and bioscience literature were quotation errors, but it is clearly unfortunate that such errors should occur in a document published by the American Psychological Association.

9.5 Conclusions

This chapter concludes Part I by summarising and discussing the key findings. Are there any differences in the legibility of serif and sans serif typefaces when they are used to generate printed material? The modal finding from the research studies that have been reviewed is that there are not. Two studies in particular have been cited in support of the superiority of serif typefaces, but these can be discounted on scientific grounds. Are there any differences in readers' preferences and connotations between serif and sans serif typefaces when they are used to generate printed material? The modal finding is that there is no overall preference between serif and sans serif typefaces, nor any overall difference in their connotations. Even so, there is a suggestion that readers' preferences and the connotations of serif and sans serif typefaces may vary between different contexts, and the chapter discussed the implications of this for educational publishing and educational assessment.

The chapter considers the relevance of the findings for previously stated assumptions about the legibility of serif and sans serif typefaces. The traditional view that "everybody knows" that serif typefaces are easier to read on paper than sans serif typefaces is clearly untenable, since this view has never been supported by sound empirical evidence. Finally, the chapter concludes by assessing the position that is adopted in the seventh (2020) edition of the American Psychological Association's *Publication Manual*. The position confounds research on reading from paper with research on reading from computer screens, and the background paper on which it depends suffers from several quotation errors (that is, reference citations that do not support the statements to which they are attached).

Open Access This chapter is licensed under the terms of the Creative Commons Attribution 4.0 International License (http://creativecommons.org/licenses/by/4.0/), which permits use, sharing, adaptation, distribution and reproduction in any medium or format, as long as you give appropriate credit to the original author(s) and the source, provide a link to the Creative Commons license and indicate if changes were made.

The images or other third party material in this chapter are included in the chapter's Creative Commons license, unless indicated otherwise in a credit line to the material. If material is not included in the chapter's Creative Commons license and your intended use is not permitted by statutory regulation or exceeds the permitted use, you will need to obtain permission directly from the copyright holder.

Part II
Reading from Screens

Chapter 10
"Everybody Knows": Reading from Screens

10.1 Introduction

The last 60 years have seen fundamental changes in the way that people acquire and share information. At the beginning of this period, information was mainly represented in the form of written text that was either read directly by readers themselves or else communicated by teachers or other speakers. Nowadays, much information is available on computer monitors or other screens. (One should, of course, recognise the "digital divide" both between people in the developed world and people in the developing world and also among diverse groups within the developed world in terms of their access to the relevant technologies.) The availability of this technology is particularly apparent in educational settings, where students routinely expect to acquire information through computer systems and also use such systems to submit their academic assignments to be evaluated by their teachers and other assessors.

Users of screen-based material have a choice in how they make use of that material. They can either read it in the form in which it is presented, or they can print it in hard copy. Shaikh (2004; Shaikh & Chaparro, 2004) carried out an online survey concerning reading habits and obtained responses from 330 participants: 120 were students, and the rest were recruited from a variety of e-mail lists. Most respondents were happy to read news items, newsletters, and product information online, but they preferred to scan articles in journals online before printing them off to read in more detail. This was consistent with the findings of previous studies that researchers and other professionals tended to print off articles or other important documents to read.

It used to be popular to assume that the increased use of digital technologies among young adults meant that they constituted a distinct population who thought and learned in qualitatively different ways from older people. They were variously called "Millenials" (Strauss & Howe, 1991), the "Net Generation" (Tapscott, 1998), "Digital Natives" (Prensky, 2001), and "Generation Y" (Jorgensen, 2003). However, these ideas were not supported by research evidence (see, e.g., Pedró, 2009). Surveys found *quantitative* differences between older and younger people in their attitudes to technology and their uses of technology, but there was no evidence for any *qualitative* differences in people born since the early 1980s (e.g., Jelfs & Richardson, 2013). In fact, many of today's students retain a strong preference for print (Baron et al., 2017; Mizrachi, 2015), although some do not (Singer & Alexander, 2017). (This may well depend on mundane factors such as local printing costs.)

Which typefaces should be used in the presentation of text on computer screens? One approach was to employ versions of typefaces that were already well established in conventional printing. The top row of Fig. 10.1 contains two examples of such typefaces. As was mentioned in Sect. 1.2, the serif typeface Times New Roman was devised in 1932 for use in the London newspaper *The Times*. Versions became popular in printing and publishing more generally, and in the 1980s these were provided by Macintosh and Microsoft in their word-processing software. The typeface known as "Times Roman" was adopted by Apple. The sans serif typeface Arial was devised in 1982 for use on IBM printers. It was adopted by Microsoft in 1990, and it was the default typeface for several years in some of its applications; it is also available for Macintosh users.

An alternative approach was to devise new typefaces, often with the aim of ensuring their legibility on small or low-resolution screens. The bottom row of Fig. 10.1 contains two examples of such typefaces. The serif typeface Georgia and the sans serif typeface Verdana were both developed for Microsoft in the 1990s and were released in 1996. They were subsequently made available for installation on Macintosh computers. Other researchers developed entire families of typefaces. In the United States, Bigelow and Holmes (1986) devised the Lucida family, while in Russia Paratype was devised for material in both Latin and Cyrillic text (Akhmadeeva et al., 2012). Other artificial typefaces were devised by Arditi (2004), Beier and Larson (2010), and Sanocki (1987), although some of these were so radically different from traditional typefaces that they might well have caused difficulties even for experienced readers.

It was mentioned in Sect. 5.1 that actual serif and sans serif typefaces typically differ in a number of characteristics. In principle, it should be possible to devise artificial typefaces in which serif and sans serif differ only in the presence or absence of serifs. In fact, efforts to devise such typefaces disclosed a major confound with the width of the letters and the inter-letter spacing. In particular, Arditi and Cho (2005) found that adding serifs to Arditi's (2004) sans serif style led to an increase in the mean width of the letters in order to accommodate the serifs. Conversely, Moret-Tatay and Perea (2011) found that removing the serifs from the serif style Lucida

Times New Roman: The quick brown fox jumps over the lazy dog.	Arial: The quick brown fox jumps over the lazy dog.
Georgia: The quick brown fox jumps over the lazy dog.	Verdana: The quick brown fox jumps over the lazy dog.

Fig. 10.1 Examples of common serif typefaces (Times New Roman and Georgia) and common sans serif typefaces (Arial and Verdana) for displaying text on screens

Bright led to an increase in the average inter-letter spacing. This confounding means that any results that are obtained using such typefaces are likely to be ambiguous.

In conventional paper-based printing, individual letters and other symbols are discrete physical entities. As was mentioned in Sect. 2.4, the overall height of typefaces (their *body size*) is traditionally expressed in terms of points, where one point is approximately equal to 0.35 mm. However, the size of typefaces is also expressed in terms of the dimensions of lowercase letters. The *x-height* of a typeface is the height of lowercase letters that do not have either ascenders or descenders (such as the letter *x* itself). When the same characters are presented on computer screens, they are simply fragmented digitised representations. Body size and x-height become relative terms since they depend on the size in which the characters are displayed on-screen. Rendering printed typefaces as digitised letter forms has been an exceedingly complex process involving complex debates and decisions (Bigelow, 2020a, b), although much of this process may well not be apparent to most display-screen users.

Section 2.5 argued that there was no reason to think that serifs and other features had the same consequences when people were reading from paper as when they were reading from computer monitors or other screens. In fact, with regard to reading from computer screens, designers and design educators have typically claimed that the legibility of sans serif typefaces was superior to that of serif typefaces (e.g., Poncelet & Proctor, 1993; Schriver, 1997, p. 508; "Universal Design", 1999, p. 5), hence my colleague's assertion, mentioned in Sect. 1.1, that "everybody knew" that sans serif typefaces were easier to read on screens than were serif typefaces. Even so, advocates of this position have usually failed to present any empirical evidence in support of this claim, and accordingly Part II of this book reviews the research literature with regard to the legibility of serif typefaces and sans serif typefaces when they are used to generate material on computer monitors and other screens.

10.2 Legibility of Serif and Sans Serif Typefaces Using Older Technology

Before considering the legibility of typefaces using modern computers, it should be acknowledged that the use of technology to enable information to be presented in other ways than as material printed on paper is by no means a new phenomenon. From the eighteenth century onwards, speakers used epidiascopes (opaque projectors also known as episcopes) and "magic lanterns" (using transparent plates) to project images of objects and other material onto viewing screens for potentially large audiences. However, from the 1950s these were superseded by overhead projectors and slide projectors. The widespread adoption of the latter technologies led to the development of recommendations for best practice. It was widely asserted, in particular, that sans serif styles rather than serif styles should be used for the projection of textual material. Nevertheless, as Phillips (1976, pp. 18–19) observed, such assertions seem to have been based on personal preferences rather than empirical research.

Adams et al. (1965) compared the legibility of different typefaces in the images produced by an overhead projector. They tested 120 children in Grades 1, 2, 3, 5, and 6 (i.e., aged 6–12) of a university's laboratory school. The children in each grade rotated among five rows of seats at varying distances from the projection screen. They were shown groups of four uppercase letters produced on an electric typewriter in five different styles and sizes and were asked to list them on a prepared response form. For one style, the letters were in a sans serif typeface (Bulletin). For the other four styles, they were produced in different sizes in a serif typeface (Elite). One of these styles, Elite 6/32 in. (4.76 mm), was the same size as the Bulletin type. Adams et al. noted that letters in the Elite 6/32 in. style were significantly more likely to be reported correctly than were letters in the Bulletin style by children in four of the five grades. Even so, they added that this "may be a phenomenon of the sample and might not similarly be observed in a replication of the study" (p. 427).

Grooters (1972) carried out a similar study in which rows of ten uppercase letters were presented to 60 adult participants in four different sizes at four different distances by means of a Kodak Carousel slide projector. Instead of actual typefaces, he employed the templates that were in use at the time for lettering in technical drawing. The participants were required to read the letters aloud as if in an eye examination. He found that the sans serif style LeRoy Standard was marginally more legible than the slab serif style LeRoy Stymie Medium, but that both of these were significantly more legible than the sans serif style LeRoy Condensed Gothic, in which the letters were 60% of the width of those in LeRoy Standard. These results were found at all distances and in all sizes, except for the closest distance and the largest size where performance approached 100% for all three styles (in other words, there were ceiling effects).

Phillips (1976) similarly compared a slab serif style of lettering (Leroy Stymie) with a sans serif style (Twentieth Century). Using a Kodak Carousel projector, he presented 31 volunteers with six slides, each containing five lines of ten randomly ordered uppercase letters in decreasing size. Three slides contained letters printed in the slab serif style, and three contained letters printed in the sans serif style, in each case using a light, medium, or bold stroke width. Once again, the participants were asked to read the letters in each slide aloud, line by line. Overall, performance was better with the slab serif style than with the sans serif style (pp. 53–54). There was however a significant interaction between the effects of letter style and letter size, such that the difference between the two styles was only significant using one of the smaller letter sizes (pp. 56–57). There was also a significant three-way interaction with stroke width, such that the difference between the two styles was only significant for two of the 15 combinations of size and stroke width (pp. 59–60). Phillips concluded that performance was so poor with the smaller letter sizes that neither style would be acceptable for projected visual materials; but that conversely performance was so good using either style with the larger letter sizes that both serif and sans serif lettering would be acceptable for projected visual materials (p. 70).

Woods et al. (2005) showed pairs of lowercase letters to groups of children from kindergarten to fourth grade using a tachistoscope attached to a slide projector. The children had to say whether the letters in each pair were the same or different and to write them down. Each pair was presented in a serif typeface (Times New Roman) or in a sans serif typeface (Arial), and the two typefaces were presented either in separate blocks of slides or in the same block. The children's scores on both discrimination and identification were higher for letters that were presented in the sans serif typeface than for those presented in the serif typeface. This was true regardless of the children's age or the size of the typeface used. One problem with this study is that the children were tested in small groups, and some cheating had taken place.

Overhead projectors and slide projectors have in turn been superseded by computer-based projection software, most obviously by Microsoft PowerPoint. (This was developed in the 1980s by an independent software company to produce both overhead transparencies and slides. However, the company was acquired by Microsoft in 1987, and it was then developed to display presentations through digital projectors on both Windows and Macintosh systems.) PowerPoint has been in general use since the early 1990s, and in practice its applications in education and in other fields have tended to borrow techniques adopted with the older technology of slide projectors. In particular, presenters are still being advised to use sans serif typefaces rather than serif typefaces in their PowerPoint presentations (e.g., Garon, 1999). More recently, Ing et al. (2017) found that 14 out of 17 speakers at an ophthalmology conference had used sans serif typefaces in their presentations. They suggested that "serif fonts may be harder to read in digital slides" (p. 172). Phillips' (1976) suggestion that such advice is based more upon personal preference than upon empirical research probably still applies, but one study has evaluated this directly.

Earnest (2003) compared the legibility of serif and sans serif typefaces for material presented using PowerPoint. He assigned 138 students to five different groups. Four groups viewed a recording of a speech given by their university's president incorporating a slide presentation, whereas the fifth group only viewed the speech. Two of the first four groups viewed slides using a serif typeface (Times New Roman), and the other two groups viewed slides using a sans serif typeface (Verdana). Immediately afterwards, the students answered nine multiple-choice questions on factual points mentioned in the speech. Nine days later, they were asked to complete the test for a second time. Earnest found that the groups who had viewed slides obtained higher scores than the group who had only viewed the speech. However, there were no significant differences between the groups who had viewed the slides in a serif typeface and the groups who had viewed the slides in a sans serif typeface.

The limited number of studies using these older technologies have failed to produce unequivocal evidence favouring either serif or sans serif typefaces. Consistent with this idea, participants tend to give similar qualitative ratings of serif and sans serif typefaces presented using either slide projectors (Kastl & Child, 1968) or PowerPoint (Mackiewicz, 2007). There is certainly no support for the notion that sans serif styles should be routinely preferred for the projection of textual material.

10.3 Issues with Screen Technology

Early computers typically lacked the facility for generating visual displays; instead, they produced text that was generated using rudimentary printers. In the 1960s, however, it was realised that visual displays might be useful, and appropriate technology was at hand to facilitate this in the form of cathode-ray tubes (CRTs), which were widely used in scientific research (as oscilloscopes) and most obviously in television sets. In CRTs, an electron gun stimulates pixels arranged in a checkerboard pattern on a phosphorescent screen, and this process is carried out repeatedly in a systematic manner to generate a visible display. Early monitors tended to be very low resolution, meaning that the details of images were lost. Indeed, this is probably the origin of the idea that sans serif typefaces should be used, because serifs would have been among the lost details. Schriver (1997, p. 403) noted that designers for television had mainly used sans serif typefaces for many years (see also McVey, 1985). Even so, technology improved, and by the year 2000 high-resolution colour CRT monitors had been developed.

In the 1990s, however, an alternative form of technology became available through liquid crystal displays (LCDs). These use liquid crystals to modulate the light emitted from a background to generate images on a computer screen, again using a checkerboard pattern of pixels that is repeatedly scanned. LCD monitors were initially developed for use with laptop computers because of their reduced size, weight, and power consumption; however, during the 2000s they became available more generally, and their resolution typically exceeded that of CRT monitors. As a result, since the late 2000s LCD monitors have generally superseded CRT monitors in computer-based applications.

One issue with the presentation of text using both CRTs and LCDs is that of aliasing. In general terms, this is the under-sampling of the information needed to produce an accurate reproduction of a particular character. More specifically, in both CRT and LCD technology, text is displayed in the form of arrays of square pixels. If a stroke in a character is oblique rather than horizontal or vertical, its contour will receive only an approximate representation as an array of pixels. As a result, it will appear irregular and ragged rather than continuous (an effect sometimes known as "staircasing" or "the jaggies"). In theory, such "aliased" text should be less legible than text printed on paper, and the effect should be more pronounced with low-resolution monitors than with high-resolution monitors.

This issue was originally handled by means of anti-aliasing software. This smoothed the edges of characters by averaging the surrounding pixels to yield varying levels of grey scale in the contours of the displayed text. Gould et al. (1987) found that text presented on paper was read significantly faster than aliased text presented in the same typeface on a CRT; however, the difference became nonsignificant when text presented on paper was compared with anti-aliased text presented in the same typeface. An analogous process known as "scale-to-grey" was used when displaying images scanned from printed text on computer monitors. Sheedy and McCarthy (1994) found that scale-to-grey led to enhanced reading performance compared to

10.3 Issues with Screen Technology

simple black-and-white text, and participants reported fewer symptoms of eye strain as a result of reading scale-to-grey text; as expected, both differences were greater with low-resolution CRTs than with high-resolution CRTs.

The increasing use of LCDs in the 1990s enabled a different approach to be taken to the aliasing issue. In LCDs, each pixel consists of three vertical bars representing the colours red, green, and blue. In 1998, Microsoft introduced ClearType software which used sub-pixel rendering with the aim of enhancing the legibility of text presented in LCDs. However, later research using a variety of tasks failed to show an unequivocal advantage of ClearType text over aliased text (Aten et al., 2002; Dillon et al., 2004, 2006; Gugerty et al., 2004; Slattery & Rayner, 2010; Tyrrell et al., 2001). Gugerty et al. (2004) also compared ClearType text with anti-aliased text using 10-point Verdana. They found that words presented in anti-aliased text were read significantly more slowly and less accurately than words presented in ClearType or aliased text, and they argued that anti-aliasing software should not be employed with smaller type sizes.

One problem with ClearType software was that the resulting characters had coloured borders that were generally thought to be distracting for readers. Microsoft therefore offered ClearType with five levels of sub-pixel rendering that varied from greyscale with no colour filtering to a high level of colour contrast. Sheedy et al. (2008) evaluated both objective performance and subjective preference for material presented in a sans serif typeface (Verdana) with all five levels of sub-pixel rendering. They found that readers preferred a moderate level of ClearType rendering to higher levels or to greyscale, but that ClearType rendering did not improve text legibility, reading speed, or reading comfort.

Another issue with screen technology is the rate at which images are refreshed: with CRTs, too slow a refresh rate leads to flicker. Wilkins (1986) showed that flicker tended to disrupt readers' saccadic eye movements, even when the refresh rate was sufficiently high to render the flicker imperceptible. (Wilkins showed that eye movements were also disrupted when readers read a printed page that was illuminated by conventional fluorescent lighting.) With LCDs, the screen itself does not flicker, but in many models the screen is backlit using pulse width modulation. The flicker may not be perceptible, but readers' eye movements may be disrupted, and they may complain of discomfort (Brown et al., 2020; Wilkins, 2021). Thus, any findings regarding participants' eye movements obtained when reading from either CRTs or LCDs need to be interpreted with caution.

10.4 Conclusions

This chapter has discussed the increased use of screen-based reading in education and in daily life generally. Readers usually have the option of printing off screen-based material to be read on paper, and this seems to be popular when researchers have to read more serious material. Both the use of computer technology and attitudes to such technology seem to vary with the user's age, but there is no support for the so-called "digital natives" hypothesis. Some existing typefaces were taken over to use in computer systems, while other typefaces were developed specifically for on-screen use. The chapter discussed the legibility of serif and sans serif typefaces when projected by means of older technology such as slide projectors, overhead projectors, and PowerPoint, but this did not show any consistent difference in their legibility. Finally, the chapter described some of the technical issues concerning the way that images are displayed using CRTs and LCDs.

Open Access This chapter is licensed under the terms of the Creative Commons Attribution 4.0 International License (http://creativecommons.org/licenses/by/4.0/), which permits use, sharing, adaptation, distribution and reproduction in any medium or format, as long as you give appropriate credit to the original author(s) and the source, provide a link to the Creative Commons license and indicate if changes were made.

The images or other third party material in this chapter are included in the chapter's Creative Commons license, unless indicated otherwise in a credit line to the material. If material is not included in the chapter's Creative Commons license and your intended use is not permitted by statutory regulation or exceeds the permitted use, you will need to obtain permission directly from the copyright holder.

Chapter 11
The Legibility of Letters and Words

11.1 Reading Letters and Words in Serif and Sans Serif Typefaces

One of the earliest uses of computers in vision research was as tachistoscopes for the brief presentation of visual stimuli. One fundamental problem was that cathode-ray tubes (CRTs) suffered from the decay time or persistence of the phosphor in the individual pixels, which led to inaccurate measurements (see Hutner et al., 1999). (It should be noted that traditional tachistoscopes were by no means immune to such problems: Mollon & Polden, 1978.) Liquid crystal displays (LCDs) are generally regarded as more appropriate for vision research (Wang & Nikolić, 2011). Even so, it is generally regarded as being good practice to use a *backward-masking* procedure in which a second stimulus or *mask* (perhaps only a random pattern) is presented after a brief interval to overwrite the visual trace of the original stimulus.

Suen and Komoda (1986) used this approach to compare the recognition of individual uppercase and lowercase letters that had been digitised using an optical scanner from print in a slab serif typeface (Courier), a sans serif typeface (Letter Gothic), and the output of a dot-matrix printer. All three typefaces were monospaced or non-proportional (that is, each character occupied the same width). The characters were presented on a high-resolution CRT screen controlled by an Apple microcomputer. (The actual duration of the presentation was not specified.) They were followed by a mask after 0 ms, 16.7 ms, or 33.3 ms. If the mask was presented immediately after the letters, performance was best with the sans serif typeface and worst with the dot-matrix style. The differences among the three styles were much reduced if the mask was presented after a delay, but this may have been a ceiling effect, since performance was 80% or better with all three styles, presumably because of the additional time available for processing the letters.

As mentioned in Sect. 4.1, Korean is another language where the alphabet can be rendered in either serif (or Ming) typefaces or sans serif (or Gothic) typefaces. Hwang et al. (1997) presented Korean participants with CRT screens consisting of 12 windows, each containing letters in one of six different sizes. The participants' task was to find and read aloud a particular target letter and then to report how much visual fatigue they had experienced while carrying out the task. The letters were presented using either an unspecified Ming typeface or an unspecified Gothic typeface. The participants' response times were faster, their accuracy was higher, and their reported visual fatigue was lower when reading letters in a sans serif typeface than when reading letters in a serif typeface. Even so, different results were obtained by Kong et al. (2011) in a study that was described in Sect. 4.1. They asked Korean participants to read aloud sets of four one- or two-syllable letters of varying sizes and to rate how much discomfort they had experienced when reading each set. The letters were presented either on paper or on an LCD screen using either an unspecified Ming typeface or an unspecified Gothic typeface. When the letters were presented on an LCD screen, there was no difference between the serif typeface and the sans serif typeface either in the participants' performance or in their reported discomfort.

As was mentioned in Sect. 5.1, Arditi (2004) devised software to generate typefaces with slab serifs of varying size. Arditi and Cho (2005) used this software to construct lowercase typefaces of uniform thickness with slab serifs extending 0% (sans serif), 5%, or 10% of the cap height (the height of capital or uppercase letters). In theory, the resulting typefaces should have varied only in the size of the serifs, but, as was mentioned in Sect. 10.1, Arditi and Cho found that an increase in the spacing between successive letters had been required to accommodate the serifs. They therefore used an inter-letter spacing of 0%, 10%, or 40% of the cap height, yielding a 3×3 design. Arditi and Cho measured size thresholds when random five-letter strings were presented on a CRT computer screen as black letters against a white background. Data were obtained from four participants with normal vision. There was a large effect of spacing such that closely spaced letters yielded higher thresholds (i.e., poorer performance). There was also a significant but small effect of serif size, such that serifs of 5% or 10% led to lower thresholds (i.e., better performance) than a sans serif typeface, which Arditi and Cho ascribed to the concomitant increase in spacing required to accommodate them.

After Microsoft introduced ClearType software with the aim of tackling the aliasing issue in text presented on LCDs (see Sect. 10.3), a range of new typefaces was commissioned to try to exploit this new technology. They included two serif typefaces (Cambria and Constantia) and four sans serif typefaces (Calibri, Candara, Corbel, and Consolas, the last for use mainly in programming). Chaparro et al. (2006a, b, 2010) evaluated the legibility of these new typefaces in comparison with that of the serif typeface Times New Roman and the sans serif typeface Verdana. Nine participants were presented with individual characters from all eight typefaces for just 34 ms each (but with no backward mask) using an LCD monitor with ClearType software enabled and were asked to say each character's name aloud. The proportion of characters reported correctly was highest for Consolas, Cambria, and Verdana and lowest for Times New Roman, Candara, and Corbel. Taking Times New Roman as the reference, accuracy was significantly better for Consolas, Cambria, and Verdana but not for the other four typefaces (Chaparro et al., 2010). Despite these somewhat ambivalent findings, Microsoft made Calibri the default typeface for all its Office applications in 2007 (and it remains the default at the time of writing).

11.1 Reading Letters and Words in Serif and Sans Serif Typefaces

Moret-Tatay and Perea (2011) used a lexical decision task in which Spanish students had to say whether or not stimuli were genuine words. Half the stimuli were Spanish words, and the other half were nonwords created by changing two letters in genuine Spanish words. Each was presented in the centre of an LCD screen in either a serif typeface (Lucida Bright) or a sans serif typeface (Lucida Sans). Once again, the typefaces should have differed only in the presence or absence of serifs, but, as mentioned in Sect. 10.1, Moret-Tatay and Perea noted that the serifs had occupied some of the space between the letters, and so removing the serifs led to a slight *increase* in the inter-letter spacing. They found that participants responded significantly more quickly to words presented in a sans serif typeface than to words presented in a serif typeface, and they suggested that this might have been due to the slight increase in inter-letter spacing.

11.2 The "Stripiness" of Words Displayed on Screens

Section 4.2 described research by Wilkins et al. (2007) which measured the vertical "stripiness" of a word by the height of the first peak of the autocorrelation between an image of the word and a second, horizontally displaced image of the same word. They had found that words with a higher first peak (i.e., more stripy words) were read more slowly than words with a lower first peak (i.e., less stripy words). However, it was not clear whether this led to variations in how quickly words in different typefaces were read.

Liversedge et al. (2006) had displayed sentences to 15 students as white letters on a black background using a CRT screen and monitored the movements of both their eyes. They found that in normal binocular reading the two eyes were often misaligned after a saccade, so that part of the duration of the subsequent fixation was taken up correcting this disparity in order to achieve binocular vergence. Jainta et al. (2010) presented 32 German participants with 120 unrelated sentences in blocks of 10 to read silently from a CRT screen. They found that the participants achieved better binocular vergence when the sentences contained words with a higher first peak, but that this took longer to achieve and led to a longer overall fixation duration. They argued that these findings explained the longer overall reading time for words with higher first peaks.

Wilkins et al. (2020) observed that different typefaces appeared to vary in the periodicity of their letters' vertical strokes. In two serif typefaces, Times and Palatino, the letter strokes were relatively evenly spaced, whereas in two sans serif typefaces, Arial and Verdana, the spaces between the strokes within a letter were greater than the spaces between the letters, leading to low periodicity. Wilkins et al. determined the first peak of the horizontal autocorrelation for passages from two novels when they were printed to the Retina (LCD) screen of an Apple Macbook Pro in each of nine serif typefaces and in each of 11 sans serif typefaces. They found that the first peak of the horizontal autocorrelation was significantly greater for the serif typefaces than for the sans serif typefaces. Wilkins et al. argued that this difference in the first peak of the horizontal autocorrelation was not due to the serifs themselves but to the effect of the serifs on the rhythm or periodicity of the typefaces. However, Wilkins et al. did not provide any evidence that these differences led to significant variations in how quickly words in different typefaces were read.

11.3 Confusions Among Letters in Serif and Sans Serif Typefaces

In their original study involving the presentation of uppercase and lowercase letters in either a slab serif typeface or a sans serif typeface (mentioned in Sect. 11.1), Suen and Komoda (1986) observed that with both typefaces errors often consisted of confusions between uppercase and lowercase forms of the same letter or confusions between letters that were visually similar in the same case. The total number of confusions was similar in the two typefaces, but there were certain differences in the pattern of confusions. For instance, with the sans serif typeface, lowercase letters tended to be mistaken as their uppercase counterparts rather than vice versa, but the reverse tended to be true for the slab serif typeface. Suen and Komoda ascribed these trends to the design of the characters in the relevant typefaces rather than to the presence or absence of serifs.

Using a laptop computer with an LCD screen, D. Fox, Chaparro, and Merkle (2007) presented individual characters (the 26 lowercase letters plus the digits 0–9 and 11 common mathematical and scientific symbols) to ten participants in ClearType rendering for just 34 ms (but with no backward mask) in each of 20 different typefaces. The participants were asked to say each character aloud and were scored on their accuracy. Fox et al. focused on errors for the letter *e* (which is confusable with the letter *c* and the number *0*) and for the number *0* (which is confusable with the letters *e* and *o*). For the letter *e*, the most accurate performance was obtained with the sans serif typefaces Clearview Text and Verdana, and the least accurate performance was obtained with the serif typeface Garamond. For the number *0*, the most accurate performance was obtained with the serif typefaces Centaur and Rockwell, and the least accurate performance was obtained with the serif typeface Constantia.

In further results from this study, Fox et al. (2008) compared the data for the numbers *0* and *1* (which is confusable with the letter *l*). For the number *1*, the most accurate performance was obtained with the sans serif typeface ClearView Text, and the least accurate performance was obtained with the serif typeface Centaur. Fox et al. employed classification tree analysis to identify the physical features of different typefaces that might be responsible for variations in legibility for both numbers, although they did not include the presence or absence of serifs as a feature in these analyses. Taken together, the findings of this study suggest that some characters tend to be more legible when presented on-screen in sans serif typefaces than when presented on-screen in serif typefaces, whereas the opposite is true for some other characters.

Beier and Larson (2010) constructed three new artificial typefaces. In each case, they varied the presence or absence of slab serifs with no other changes to the letters themselves. In a pre-test, each of 34 participants viewed the letter *d* presented on the LCD screen of a laptop computer that had been placed on a podium at eye-level height at a distance of 10 m. They approached the podium until they could correctly identify the letter. The relevant distance was then used for the presentation of individual uppercase and lowercase letters in the main experiment. Beier and Larson found that serifs could serve either to enhance or to impair the relative differentiation of individual letters. For instance, a slab serif added to the top of the stem of the letter *i* led to improved identification, but this was not the case when slab serifs were added to both the top of the stem and the baseline. In Sect. 4.2, it was noted that similar findings had been obtained with the identification of individual characters when reading from print (Harris, 1973; Tinker, 1963, p. 36). However, Vernon's (1929) findings, again based on reading from print, imply that such confusions would be much less likely if letters were presented in the context of meaningful text, as in normal reading.

11.4 Conclusions

As with reading from print, the earliest research on the legibility of different typefaces when reading from screens was concerned with recognising individual letters and words under different conditions. Studies that employed authentic typefaces showed at most that *some* sans serif typefaces are more legible than *some* serif typefaces. Research using artificial typefaces suffers from confounding between (a) the presence or absence of serifs and (b) variations in the width of the letters and the spacing among successive letters. The horizontal autocorrelation of individual words differs across different typefaces, but it is not clear whether this leads to differences in how quickly different typefaces can be read. The vertical stripiness of serif typefaces tends to be greater than that of sans serif typefaces, but there is little evidence that this leads to variations in how quickly words in different typefaces are read. As with printed letters and words, identification errors are often the result of confusions among visually similar letters, but visual confusions are not more likely with sans serif typefaces than with serif typefaces, contradicting an old hypothesis that serifs make letters easier to discriminate (Legros, 1922, p. 11).

Open Access This chapter is licensed under the terms of the Creative Commons Attribution 4.0 International License (http://creativecommons.org/licenses/by/4.0/), which permits use, sharing, adaptation, distribution and reproduction in any medium or format, as long as you give appropriate credit to the original author(s) and the source, provide a link to the Creative Commons license and indicate if changes were made.

The images or other third party material in this chapter are included in the chapter's Creative Commons license, unless indicated otherwise in a credit line to the material. If material is not included in the chapter's Creative Commons license and your intended use is not permitted by statutory regulation or exceeds the permitted use, you will need to obtain permission directly from the copyright holder.

Chapter 12
Reading and Comprehending Text

12.1 Reading Text in Serif and Sans Serif Typefaces

The work by Gould et al. (1987) was mentioned in Sect. 10.3. In another experiment, Gould et al. followed a number of previous studies comparing screen and paper presentation in a proof-reading task: 18 research workers read through articles of about 1,000 words to try to locate between 1 and 10 misspelled words by saying them aloud to the experimenter. The screen versions were presented as antialiased characters on a cathode-ray tube (CRT) screen using a system which had been specifically designed to simulate their appearance on paper, and high-quality printed versions were generated using the same script files. The screen and paper versions of each article were both presented using a serif typeface (Press) and two sans serif typefaces (Letter Gothic and Univers). Each participant read one article on-screen and one article on paper in each of the three typefaces. Overall, the paper versions were read significantly faster than the screen versions, although this might have been partly because the screen system required 1.5 s scrolling time to go from one page of text to the next. There was no significant difference in the participants' accuracy between the two versions. There was no significant variation across the typefaces and no significant interaction between the effects of display mode and typeface in either speed or accuracy.

Also using a CRT screen, Tullis et al. (1995) compared the legibility of four typefaces available in the Microsoft (MS) Windows operating environment: two serif typefaces, MS Serif and Small Font, and two sans serif typefaces, Arial and MS Sans Serif. Each was used in two, three, or four different sizes, yielding 12 combinations of typeface and size. Each of these was shown in either a bold or regular style and against a white or grey background, resulting in a total of 48 conditions that were administered in a random sequence. All of the participants viewed the same paragraph of text presented in each condition and were asked to count the number of typographical errors that it contained. After they had read the paragraph, they pressed the "Enter" key and reported the number of errors using a dialogue box. Tullis et al. measured the time taken to read the paragraph and whether or not the correct number of errors had been reported. The two sans serif typefaces yielded faster reading times, higher accuracy, and higher ratings than the two serif typefaces, especially with larger type sizes.

Garcia and Caldera (1996) asked ten students to read aloud paragraphs of 30 words from a computer screen (neither the computer nor the monitor was specified). The paragraphs were presented in three different type sizes and in five different typefaces: a serif typeface (Times New Roman), three sans serif typefaces (Arial, System, and MS Sans Serif), and a cursive typeface (i.e., a typeface intended to mimic handwriting) (Lucida Casual). There was significant variation across the 15 conditions, with 10-point Arial yielding the fastest reading time. System yielded the second fastest time, but no further results were reported.

Stone et al. (1999) asked 48 female survey interviewers to read aloud 24 sets of material, each consisting of 30 random words chosen to be at the eighth-grade reading level. Half of the sets were presented in black ink on white paper, whereas the other half were presented in black type against a white background on the liquid crystal display (LCD) screen of a laptop computer to simulate the former. In both cases, each set of words was presented on a single page, and the order of the conditions was counterbalanced. The sets of words were presented in three different typefaces: a sans serif typeface (Helvetica), a serif typeface (Times Roman), and a slab serif typeface (Courier). Stone et al. found that their participants read the sets in serif typefaces faster than the sets in sans serif typeface, but there was no significant variation in the number of errors made. There was no significant difference in either reading speed or accuracy between the sets presented on paper and those presented on-screen, and neither of the interactions with the effects of typeface and mode of presentation was significant.

Josephson (2008) carried out an exploratory study of eye-movements in reading from screens. She presented six participants with four news stories of around 250 words on a high-resolution monitor, and their eye-movements were tracked using a corneal reflection system. The stories were presented in the four different typefaces shown in Fig. 10.1 in Sect. 10.1. Afterwards, the participants were asked to rate each typeface on several 10-point scales and then to say which was easiest to read and which they liked the most. The story presented in Verdana was read the fastest, followed by that presented in Times New Roman. The story presented in Times New Roman yielded the fewest fixations, whereas that presented in Verdana yielded the most. Conversely, the story presented in Verdana yielded the fewest backward movements to reread previously presented words, whereas that presented in Times New Roman yielded the most. Verdana was rated highest whereas Times New Roman was rated lowest of the four typefaces. When asked which typeface was the easiest to read, three of the participants chose Verdana, whereas none chose Times New Roman. Apart from the small number of participants, one limitation of this study was that one story was assigned to each typeface, and so differences among the typefaces were directly confounded with differences among the news stories themselves.

Banerjee and Bhattacharyya (2011) presented 40 young postgraduate researchers with 18 passages of roughly the same length on an LCD monitor. The passages were presented in one of three serif typefaces (Times New Roman, Georgia, and the slab serif Courier New) or one of three sans serif typefaces (Arial, Tahoma, and Verdana) in one of three sizes (10, 12, or 14 points). The passages were assigned at random to the 18 conditions for each participant and presented in a different random order to each participant. The time taken by the participants to read each passage was recorded, they then rated the overall ease or difficulty of reading each passage, and finally they completed a short questionnaire on the mental workload involved in reading each passage. Their eye movements were also tracked using a binocular eye movement recorder.

There was a significant interaction between the effects of typeface and type size on the participants' reading time. The effect of type size was significant for Courier New and Arial, but not for the other four typefaces. The mean reading time was significantly less for the serif typefaces than the sans serif typefaces, but only for 10-point and 14-point sizes. Courier New and Arial also yielded the fastest average reading time. Verdana in 14-point was rated as easier to read than any other combinations of typeface and size, and Arial in 14-point was rated as the second most preferred. Verdana in 14-point was also rated as having the least mental workload, followed by Arial in 14-point and Tahoma in 14-point. Courier New yielded the lowest fixation duration, followed by Verdana and Arial; Courier New also yielded the lowest total gaze duration, followed by Verdana. In contrast, Times New Roman yielded the longest fixation duration and the longest total gaze duration. However, there were no overall differences between serif and sans serif typefaces in eye movement parameters.

Perea (2013) presented 24 Spanish undergraduate students with individual sentences on a computer screen and measured their eye movements using a video-based eye-tracking device. Each sentence appeared when the participant looked at a square on the left-hand side of the screen, and the participant pressed a key when they had finished reading the sentence to themselves. To check on their comprehension, a yes/no question was presented after 20% of the sentences. (Overall, 96% of these questions were answered correctly.) The sentences were presented in four blocks of 30. Half were presented in the serif typeface Lucida, whereas half were presented in the sans serif typeface Lucida Sans. There was no significant difference in the participants' reading times, in the total number of fixations, or in the mean duration of their fixations between the serif and sans serif sentences. Perea concluded that the presence or absence of serifs did not materially affect the process of normal reading, although different results might have been obtained with longer extracts, as in Josephson's study.

12.2 Comprehending Text in Serif and Sans Serif Typefaces

As with research on reading from paper (see Chap. 5), asking the participants to read continuous text provides less opportunity for researchers to impose experimental control over their reading behaviour. Some researchers have therefore focused on their participants' *comprehension* of meaningful material read from screens rather than upon its legibility per se.

Williams (1990) carried out an experiment in which 56 students read two passages on a CRT monitor and answered four comprehension questions after each passage. Each passage contained 650–700 words and filled seven screens, so that the students had to scroll down to read the complete passages. The material was presented in either a serif typeface (10-point Times Roman) or a sans serif typeface (9-point Helvetica); the use of different type sizes ensured that the typefaces were of similar x-height. The participants were randomly assigned to five different groups. Four groups saw the two passages in different typefaces, with the order of the two passages and the order of the two typefaces counterbalanced across the four groups. The fifth group saw both passages in the sans serif typeface to check for any difference in difficulty between the two passages. The monitor screen also contained a clock, and the students were asked to record the times when they started and finished reading each passage. The comprehension questions were multiple-choice with five alternatives. The fifth (control) group showed no difference in mean reading rate between the two passages. Across the other four groups, the mean reading rate was 14.75 words/min faster for the sans serif typeface than for the serif typeface, but the difference was not statistically significant.

Lenze (1991) used a personal computer to present 84 students with a brief paragraph and then asked them a question based on its content to which they would have to type a one-word answer. The paragraph was initially presented for 1 s; if the participant answered incorrectly, the paragraph was presented for 1 s longer, and this process was continued until they answered the question correctly. They were then presented with three further paragraphs in the same way. Half the participants chosen at random were shown the paragraph in a serif typeface, and the other half were shown the paragraph in a sans serif typeface. (Neither of the typefaces was specified.) Finally, all the participants were shown examples of text in both serif and sans serif typefaces and were asked which they preferred. Lenze found that there was no sign of any significant difference between the two groups of participants in the time needed to achieve comprehension of the texts, which suggested that serif typefaces and sans serif typefaces were equally effective in supporting reading comprehension. Even so, 77% of the students preferred the sans serif typeface to the serif typeface.

Boyarski et al. (1998) compared the serif typeface Georgia and the sans serif typeface Verdana, both of which had been designed for on-screen display. Sixteen university staff and graduate students read two passages from a standard reading test presented in the two typefaces in the same body size in Microsoft Word without anti-aliasing. The orders of the typefaces and the passages were both counterbalanced. They were allowed 1 min to read each passage and were then asked four questions to test their comprehension of the passage (each answer being scored between 0 and 3). Finally, the participants were asked to compare the two typefaces. A measure of "effective reading speed" was obtained by dividing their comprehension score by their actual reading time on each passage in order to allow for the possibility of a trade-off between speed and accuracy. Their comprehension scores were significantly higher for passages in Georgia than for passages in Verdana. Nevertheless, there was no significant difference between the two conditions either in their actual reading time or in their effective reading speed. The participants judged the passage presented in Verdana to be easier to read than the passage presented in Georgia, but there was no significant difference in their ratings of the passages' sharpness or legibility.

Hojjati and Muniandy (2014) presented 30 international students at a Malaysian university with four 200-word English texts in different typefaces. Two were presented in the serif typeface Times New Roman, and two were presented in the sans serif typeface Verdana; in each case, one text was presented single-spaced, and the other was presented double-spaced. After each text, the students were presented with questions designed to test their retention of its content. Finally, they were asked to rate the ease with which they had been able to read each text on a 6-point scale. Regardless of spacing, they found text displayed in Verdana easier to read than text displayed in Times New Roman, they read text displayed in Verdana more quickly, and they recalled more of the content of text displayed in Verdana. However, the researchers had used Amazon Kindles (which use microcapsules containing electronic "ink") rather than LCDs, and at least some of the texts did not fit completely onto the visible area of the screens (p. 167). The material had been taken from Wikipedia and other sources and had been checked by subject experts, but the texts themselves suffered from grammatical and stylistic problems. It is also not clear whether the typefaces used for different texts were counterbalanced across participants or whether each text was always presented in the same typeface.

Csilla et al. (2016) asked 74 Hungarian students to read to themselves four self-contained excerpts from a Hungarian translation of *The Nature of Space and Time* by Hawking and Penrose (1999). The excerpts ran for four pages in the original book but had been reformatted to fit onto two pages of European A4-sized paper. Each excerpt was prepared in six different typefaces and then saved in Portable Document Format. Two excerpts were presented on paper, and two were presented on an LCD computer monitor. In each case, one excerpt was presented in a serif typeface randomly chosen from Book Antiqua, Garamond, and Times New Roman, and the other was presented in a sans serif typeface randomly chosen from Arial, Calibri, and Verdana, so that each student saw four different typefaces. The order of the four texts was randomised for each student. The students timed themselves reading each excerpt and then answered a number of questions about its content.

Csilla et al. analysed the data for each of the four excerpts separately and carried out independent-groups tests comparing (a) the participants who had read the excerpt on paper with those who had read the excerpt on-screen and (b) the participants who had read the excerpt in a serif typeface with those who had read the excerpt in a sans serif typeface. There were no significant differences in either reading time or comprehension for any of the four excerpts. It is unfortunate that Csilla et al. did not increase the statistical power of their analysis by using the data from all four excerpts and carrying out repeated-measures tests on the variables of presentation medium and typeface within the 74 participants. They also did not examine whether there was any interaction between the effects of these two variables on either reading time or comprehension.

12.3 Rapid Serial Visual Presentation

The introduction of new technologies that enable images to be presented in a variety of visual forms prompted new research methods to be used for investigating reading. One such method is that of rapid serial visual presentation (RSVP), in which letters, words, or groups of words are presented one at a time at the reader's point of fixation. This method was first developed by Gilbert (1959a, b), who presented sentences containing eight words using a movie projector to each of 76 participants and asked them to write down immediately afterwards what they had seen. He found that they were much more accurate when successive pairs of words were shown at the point of fixation than when successive pairs of words were presented across the screen as if in a line of text. Gilbert concluded that the former procedure enhanced the identification of stimuli by eliminating the need for eye movements.

Rubin and Turano (1992) digitised characters from the output of a laser printer using a Times Roman typeface and then used them in text displayed one word at a time on a high-resolution screen controlled by an IBM computer. They found that readers who were previously unfamiliar with this procedure could achieve reading speeds that were several times faster than when reading the same material presented on-screen as a single paragraph, typically in excess of 1,000 words/min for reading aloud or as fast as 1,800 words/min for silent reading.

As was mentioned in Sect. 11.1, Suen and Komoda (1986) had digitised printed characters in a slab serif typeface (Courier), a sans serif typeface (Letter Gothic), and the output of a dot-matrix printer. They used these characters to present paragraphs of roughly 160 words drawn from magazine articles to 36 participants. Each paragraph was shown on a high-resolution CRT screen, one word at a time for just 100 ms. Multiple-choice questions (how many was not specified) were then employed to assess the participants' comprehension of the paragraph. The dot-matrix style yielded the worst performance, but there was no difference between the comprehension of material presented in the slab serif typeface and that presented in the sans serif typeface. Suen and Komoda argued that the higher-order skills and strategies involved in reading connected text had overridden any differences in legibility between the serif and sans serif typefaces.

Yager et al. (1998) used the RSVP method to compare the legibility of two typefaces matched for x-height: Dutch, a serif typeface similar to Times; and Swiss, a sans serif typeface similar to Helvetica. Sentences were presented on a CRT monitor in white letters on a black background under conditions of either high luminance or low luminance. The latter was intended to stress the visual system and hence to simulate the situation of readers with visual impairment. The participants were 46 normally sighted college and high-school students who were asked to read each sentence aloud immediately after it had been presented. The presentation rate was calibrated to identify the threshold for correct responding for each participant. Under high luminance, there was no significant difference between the number of sentences read correctly in the two typefaces. Under low luminance, performance was significantly better with the Swiss typeface. Yager et al. noted that in this situation the Dutch typeface was close to the threshold of visual acuity. They speculated that under conditions of low luminance either the serifs or the thinner strokes of the letters of the Dutch typeface tended to be invisible, and that this was responsible for the significantly poorer reading performance.

Also using RSVP, Morris et al. (2002) presented 27 participants with words that made up meaningful but unconnected sentences. The words were presented on a CRT monitor in slab serif and sans serif typefaces taken from Bigelow and Holmes' (1986) Lucida styles to ensure that the typefaces were matched in all respects except for the presence or absence of serifs. When the words were presented at the equivalent of 14-point type (a normal reading size), there was no difference in the number that were read correctly in the slab serif and sans serif typefaces. However, when they were presented at the equivalent of 4-point type (a small but still tolerable size), performance was better with the sans serif typeface than with the slab serif typeface. Morris et al. suggested that the letters appeared relatively crowded in the latter situation, and that rendering serifs in small sizes might be counterproductive.

As mentioned in Sect. 5.1, Arditi and Cho (2005) devised lowercase typefaces of uniform thickness with slab serifs that extended for 0% (sans serif), 5%, or 10% of their cap height. They presented sentences in these typefaces on a CRT display screen using RSVP and adjusted the presentation speed for each participant to ensure a 50% correct reading rate. Data were obtained from two participants with normal vision and two with impaired vision. The former participants achieved higher reading speeds than the latter participants, but there was no effect of serif size and hence no effect of the presence versus the absence of serifs. Arditi and Cho acknowledged that the small number of participants was a limitation of their study.

One motivation for investigating the RSVP procedure was that it was suspected it might help to compensate for the limitations of handheld devices with small screens, such as cellular phones, palmtop computers, and personal digital assistants (Bernard, Chaparro, & Russell, 2000; De Bruijn et al., 2002). Palmtop computers and personal digital assistants were fairly popular towards the end of the twentieth century, but during the 2000s their functionality was superseded by that of smartphones. It is thus not surprising that the RSVP procedure also became less popular as a research tool. In any case, a major criticism of the procedure is that it lacks ecological validity, insofar as it does not represent a situation that is characteristic of normal reading in real-life settings (Perea, 2013).

12.4 Reading Material on Handheld Devices and Smartphones

It has tended to be assumed that sans serif typefaces are more legible than serif typefaces when used on handheld devices or smartphones. The sans serif typefaces Droid and Roboto were developed for Android mobile phones, while the sans serif typeface San Francisco was devised for Apple products, although all three typefaces are available in serif styles. The limited amount of research comparing the legibility of serif and sans serif typefaces when employed on handheld devices and smartphones has been carried out by Korean researchers.

Park et al. (2008) presented texts to four Korean students by means of a personal digital assistant (a Pocket PC 2002). They were asked to read the texts silently, and their eye movements were monitored using an eyeball-tracking camera which reflected infrared rays on their corneas. The texts were shown in three type sizes and in three typefaces: two serif typefaces (Batang and Gungseo) and one sans serif typeface (Gulim). The participants' fatigue was measured both by monitoring their blink rate and by asking them to rate how easy or difficult it had been to read each text on a 7-point scale. There was no significant variation across the three typefaces in their reading speed, their error rate, their eye-blinking, or their subjective ratings.

Kim et al. (2015) presented 14 Korean students with pairs of two-syllable words using an Apple iPhone with a display of 90 mm by 50 mm using one of two serif typefaces (Batang or Gungseo) or one of two sans serif typefaces (Dodum or Gulim). There were 10 trials for each typeface, and the order of the typefaces was counter-balanced across the participants. On each trial, one member of the word pair had been designated as the target, whereas the other was a distractor, and the participants' task was to read aloud the target in each pair. Their mean reading time was marginally faster for the words shown in serif typefaces than for words shown in sans serif typefaces, but the difference between the two means was not at all statistically significant.

12.5 Connotations of Serif and Sans Serif Typefaces

Several of the studies mentioned asked participants to express a preference between serif and sans serif typefaces. Misanchuk (1989) argued that, in the absence of evidence for objective differences in legibility between typefaces, designers might be guided by readers' preferences or satisfaction ratings. In fact, several studies found a significant preference for sans serif typefaces over serif typefaces (Boyarski et al., 1998; Hojjati & Muniandy, 2014; Lenze, 1991; Tullis et al., 1995), some found no significant difference (Garcia & Caldera, 1996; Holleran, 1992; Muter & Mauretto, 1991), but none found a significant preference for serif typefaces over sans serif typefaces.

Savory et al. (2012) found a clear preference for sans serif typefaces among military experts in the highly specialised area of the design of radiation detector screens. They used a focus group to discuss other aspects of these devices, a methodology where responses can be vulnerable to peer pressure from the other participants. However, the participants expressed their preferences among the different typefaces in an individual written survey, which should have avoided this problem.

Section 5.4 noted research that had focused on the connotations of different typefaces when used for print-based text, and analogous research has been carried out on the connotations of different typefaces when presented on computer monitors. Shaikh

12.5 Connotations of Serif and Sans Serif Typefaces

et al. (2006) examined the connotations of 20 different typefaces; they included four serif and six sans serif typefaces, among which were the six typefaces introduced by Microsoft to make use of ClearType software (see Sect. 10.3). Using an online survey, more than 500 participants rated the characteristics of each typeface using 15 bipolar scales and then indicated whether they would use the typeface for each of 25 online purposes.

The serif typefaces were regarded as being stable, practical, mature, and formal, but the sans serif typefaces were not regarded as being especially high or low on any of the 15 traits. The serif typefaces were regarded as being appropriate for business documents, website text, or online magazines but not for digital scrapbooking, children's documents, or e-greetings. The sans serif typefaces were regarded as being appropriate for website text, e-mail, or online magazines but not for scrapbooking, computer programming, or mathematical documents. Shaikh et al. concluded that computer users consistently attributed personalities to typefaces displayed on-screen and that both serif typefaces and sans serif typefaces were seen as being appropriate for the kinds of material that were typically read on-screen.

Shaikh (2007, pp. 44–100) carried out another online survey in which 379 participants rated text samples of 40 typefaces on 15 bipolar constructs. The typefaces included ten examples of each of four categories: serif, sans serif, display (used in advertisements or logos), and cursive (akin to handwriting). Each participant was shown 20 typefaces including five serif typefaces, five sans serif typefaces, five display typefaces, and five cursive typefaces; these were randomly selected and presented in a random order. The participants also rated the legibility of each typeface using a 7-point scale. There were significant differences between the serif and the sans serif typefaces on five of the scales: the serif typefaces were regarded as more delicate, beautiful, expensive, warm, and old, whereas the sans serif typefaces were regarded as more rugged, ugly, cheap, cool, and young. However, these differences were small in magnitude, and in general the serif and sans serif typefaces tended to be regarded as similar. The participants' ratings of legibility showed no significant difference between the serif and the sans serif typefaces, but both were rated as being significantly more legible than the display typefaces or the cursive typefaces.

Koch (2012) asked 100 volunteers to evaluate six typefaces in an online survey: five were variants of a sans serif typeface, Helvetica, and one was a slab serif typeface, Glypha Medium. For each typeface, participants were presented with the uppercase and lowercase alphabets, together with a grid showing 12 cartoon characters, each representing a different emotion (although the emotions themselves were not named). When they clicked on each character, they were shown a short animation in which the character enacted the relevant emotion; they were also shown a 5-point scale and indicated how much the relevant typeface had aroused the emotion in them (from "I do not feel this" to "I feel this strongly"). Out of the 100 volunteers, 42 completed the survey (yielding 12 ratings for each of six typefaces), of whom 32 had had some prior training in design. Comparisons between the overall ratings given to the slab serif typeface and the sans serif typefaces yielded just one significant comparison, in that Glypha Medium yielded higher ratings of satisfaction than the average across the different variants of Helvetica. However, Koch had carried out 48 different comparisons in total, suggesting that even this effect might well have been a spurious result due to chance variation (i.e., a Type I error).

Kaspar et al. (2015) carried out an online survey in which 188 students evaluated three scientific abstracts on six dimensions. The students were randomly assigned to receive the abstracts and the rating scales either in the serif typeface Lucida Bright or in the sans serif typeface Lucida Sans. (As mentioned in Sect. 11.2, these typefaces are identical except for the presence or absence of serifs.) There was no significant difference in the time taken to read the abstracts or in the time taken to complete the rating scales between the groups who received material in the two typefaces. The students who read the abstracts in the serif typeface rated them significantly more positively than the students who read them in the sans serif typeface: they rated them as significantly more comprehensible and more appealing overall, and they were significantly more interested in reading the full paper from which the abstract had been taken; they also rated the topicality, the quality, and the importance of the research question significantly more highly. Kaspar et al. noted that reports of scientific work were more often presented in a serif typeface, and they concluded that the use of such a typeface increased the impression of a work's scientific character.

Kaspar et al. (2015) carried out a similar study in which 187 students were randomly assigned to receive abstracts and rating scales either in the serif typeface Times New Roman or in the sans serif typeface Arial. (These typefaces differ on several other characteristics as well as in the present or absence of serifs.) Once again, there was no significant difference in reading speed between the groups who received material in the two typefaces. However, in this case, the participants who read the abstracts in the sans serif typeface rated them significantly more positively than the students who read them in the serif typeface: they rated them as significantly more comprehensible and more appealing overall, and they were significantly more interested in reading the full paper from which the abstract had been taken; they also rated the quality of the research question significantly more highly, although there was no significant difference between the students who read the abstracts in different typefaces in their ratings of either the topicality or the importance of the research question. Kaspar et al. concluded that there was no simple rule of thumb that favoured the presence or absence of serifs under all circumstances and that sans serif typefaces could lead to more desirable text evaluations when other features of the text compensated for the missing serifs.

12.6 Conclusions

Studies of the legibility of connected sentences that have measured readers' speed and accuracy have proved inconclusive. Research on reading text presented on computer screens has enabled investigators to use other forms of technology such as eye-tracking equipment. However, research into readers' eye movements has not yielded conclusive findings with regard to the presence or absence of serifs. As with research

12.6 Conclusions

on reading from paper, asking participants to read continuous text provides less opportunity for researchers to impose experimental control over their reading behaviour. Some researchers have instead focused on their participants' comprehension of material. Such studies have not yielded consistent findings with regard to the presence or absence of serifs on readers' speed or accuracy, and at least one study suffered from serious methodological problems.

A particular device that has been investigated is the presentation of letters, words, or groups of words one at a time at the reader's point of fixation: RSVP. This was originally thought to compensate for the limitations of handheld devices. Five studies have compared readers' comprehension, speed, or accuracy and found no significant differences except that serif styles were less legible with very small type or under conditions of low luminance when, of course, serifs are likely to have been faint or even completely invisible. It has tended to be assumed that sans serif typefaces are more legible than serif typefaces when used on handheld devices or smartphones. However, two studies failed to find any difference between serif and sans serif typefaces in terms of the participants' reading performance when using such devices.

Finally, this chapter described research on the connotations of different typefaces when presented on computer screens. Some studies, but not all, have found that readers express a preference for sans serif typefaces when reading on computer screens, but in general both serif and sans serif typefaces are regarded by users as appropriate for online purposes. There is some evidence that serif and sans serif typefaces differ in their connotations or "personality". These differences seem to reflect variations in readers' expectations, which in turn depend on their prior experience and familiarity with different typefaces.

Open Access This chapter is licensed under the terms of the Creative Commons Attribution 4.0 International License (http://creativecommons.org/licenses/by/4.0/), which permits use, sharing, adaptation, distribution and reproduction in any medium or format, as long as you give appropriate credit to the original author(s) and the source, provide a link to the Creative Commons license and indicate if changes were made.

The images or other third party material in this chapter are included in the chapter's Creative Commons license, unless indicated otherwise in a credit line to the material. If material is not included in the chapter's Creative Commons license and your intended use is not permitted by statutory regulation or exceeds the permitted use, you will need to obtain permission directly from the copyright holder.

Chapter 13
Readers with Disabilities

13.1 Readers with Visual Impairment

Plass and Yager (1995) tested 61 patients with visual impairment due to a variety of causes. Almost all preferred to read text shown on a computer monitor rather than printed on paper. Their reading rates were significantly faster when the individual words were presented on a screen using the RSVP procedure than when they were presented as a single paragraph. Although they had presented the text using both the serif typeface Times Roman and the sans serif typeface Arial, they found no significant differences in reading speed between the two typefaces (see Yager et al., 1998). Four of their patients were adults with a history of congenital nystagmus; as mentioned in Sect. 8.2, this causes visual impairment by disrupting the normal pattern of eye movements. A follow-up study by Aquilante et al. (1997) confirmed that these four patients tended to read faster with RSVP than when reading the same material as a single paragraph of text. This suggested that techniques which eliminated the need for eye movements might be useful for improving the reading ability of people with nystagmus or other visual disorders. However, Aquilante et al. did not compare their patients' performance using different typefaces.

13.2 Readers with Dyslexia

Research on people with dyslexia reading from print was described in Sect. 8.7. It was noted that the British Dyslexia Association (2018) had for many years recommended that materials for people with dyslexia should use sans serif typefaces, and this advice seems intended to apply both to material shown on screens and to material printed on paper.

Rello and Baeza-Yates (2013) recruited 48 participants between the ages of 11 and 50 who had been clinically certified as dyslexic and asked them to read 12 extracts of 60 words from a Spanish novel. These were presented in 12 different typefaces, including three serif typefaces (Computer Modern Unicode, Garamond, and Times) and four sans serif typefaces (Arial, Helvetica, Myriad, and Verdana). Assignment of the different typefaces to different extracts was counterbalanced across different participants. The extracts were presented on a liquid crystal display (LCD) screen in 14-point size, and the readers' eye movements were monitored using a corneal reflection system, both to record the duration of their fixations and to record their total reading time. After each extract, the participants answered a multiple-choice question to check their comprehension of the text, and at the end of the experiment they rated their preference for each typeface on a 5-point scale.

There was significant variation among the 12 typefaces in the participants' reading time, but there was no overall difference between the serif typefaces and the sans serif typefaces in this regard. There was significant variation among the 12 typefaces in the participants' fixation duration, and this was slightly longer for the serif typefaces than for the sans serif typefaces. There was significant variation among the 12 typefaces in the participants' preference ratings, but there was no overall difference between the serif typefaces and the sans serif typefaces in this regard. Rello and Baeza-Yates argued that shorter fixations reflected a reduced processing load and hence greater legibility, and they concluded that sans serif typefaces enhanced reading performance, even though this was not reflected in an enhanced reading time.

However, Rello and Baeza-Yates had not controlled the spacing of their different typefaces. It was noted in Sect. 8.6 that readers who are dyslexic benefit from increased spacing when reading from paper, and that this explains differences in their performance across different typefaces. It was also noted in Sect. 10.1 that the addition of serifs leads to an increase in inter-letter spacing, at least when letters are presented on screen. Perea et al. (2012) found that a small increase in inter-letter spacing could lead to enhanced performance when reading from computer screens, especially in children with dyslexia. Consequently, any suggestion in Rello and Baeza-Yates' results in favour of sans serif typefaces is likely to be due to the confounding of typeface with inter-letter spacing. Otherwise, the results of their study indicate no difference in legibility between serif and sans serif typefaces in dyslexic readers, and once again this contradicts the advice of the British Dyslexia Association (2018).

13.3 Readers with Age-Related Macular Degeneration

Several studies have investigated people with age-related macular degeneration (AMD). As was mentioned in Sect. 8.2, this causes impaired vision in the centre of the visual field. Section 11.1 described a study in which Arditi and Cho (2005) measured size thresholds when random five-letter strings were presented on a cathode-ray tube (CRT) screen as black letters against a white background. They used software to construct lowercase typefaces of uniform thickness with slab serifs extending 0% (sans serif), 5%, or 10% of the cap height (the height of capital or uppercase letters). They also used an inter-letter spacing of 0%, 10%, or 40% of the cap height, yielding

13.3 Readers with Age-Related Macular Degeneration

a 3 × 3 design. In addition to four participants with normal vision, Arditi and Cho also tested two people with AMD. They normalised the data to each participant's best performance by dividing each data point by the participant's minimum threshold. Both the participants with normal vision and the participants with AMD showed the same pattern of performance. There was a large effect of spacing such that closely spaced letters yielded higher thresholds (i.e., poorer performance). There was also a significant but small effect of serif size, such that serifs of 5% or 10% led to lower thresholds (i.e., better performance) than a sans serif typeface, which Arditi and Cho ascribed to the concomitant increase in spacing required to accommodate them. There was no evidence that the presence or absence of serifs made any difference to the performance of the participants with AMD.

Subsequent researchers tried to devise typefaces that might be helpful for people with AMD. Bernard et al. (2016) developed Eido, a monospaced sans serif typeface that emphasised the distinctive shapes of different letters. They presented normally sighted volunteers with letters, words, and sentences in either Eido or the monospaced slab serif typeface Courier using a CRT screen. They carried out six different experiments with varying print sizes. In each case, the participants used their dominant or preferred eyes, but an eye-tracking system superimposed a mask over the centre of their visual field to simulate AMD. The participants made fewer errors when stimuli were presented in Eido than when presented in Courier, but there was no difference in their reading speed between the two typefaces.

Xiong et al. (2018) used both Eido and Maxular Rx, a proportionally spaced slab serif typeface developed by Steven Skaggs that employed extra spacing between successive letters and lines. For comparison, they also used Courier, Helvetica, and Times Roman. (Helvetica is a proportionally spaced sans serif typeface, whereas Times Roman is a proportionally spaced serif typeface.) Individual sentences were presented on an LCD screen controlled by a Macintosh computer. The participants consisted of 19 individuals with AMD, 14 age-matched individuals with normal vision, and 26 young adults with normal vision. The researchers measured their fastest speed to read the sentences aloud, the smallest print size to achieve that speed, and the smallest print size that could just be read. The reading speed varied significantly across the five typefaces for the individuals with AMD, but not for the control participants. All three groups showed significant variations in the measures of print size. However, there were no systematic differences between the three serif typefaces (Courier, Maxular Rx, and Times Roman) and the two sans serif typefaces (Eido and Helvetica).

13.4 Conclusions

As mentioned in Chap. 8, any differences in the legibility of serif and sans serif typefaces might become more apparent in readers whose visual systems are challenged as the result of disablement. One study evaluated a heterogeneous sample of patients with visual impairment and found no difference in reading speed between a serif typeface and a sans serif typeface, regardless of whether text was presented on screen as a single paragraph or using the RSVP procedure. Another study evaluated a large sample of readers with dyslexia. This too found no difference in their reading speed between serif and sans serif typefaces. Any differences in the participants' preferences or in their eye movements could be attributed to the researchers' failure to control the inter-letter spacing of the different typefaces. Differences in inter-letter spacing also explain differences in reading speed in a study which simulated AMD in readers with normal vision. A study which compared reading in people with and without AMD found no systematic differences between serif typefaces and sans serif typefaces.

Open Access This chapter is licensed under the terms of the Creative Commons Attribution 4.0 International License (http://creativecommons.org/licenses/by/4.0/), which permits use, sharing, adaptation, distribution and reproduction in any medium or format, as long as you give appropriate credit to the original author(s) and the source, provide a link to the Creative Commons license and indicate if changes were made.

The images or other third party material in this chapter are included in the chapter's Creative Commons license, unless indicated otherwise in a credit line to the material. If material is not included in the chapter's Creative Commons license and your intended use is not permitted by statutory regulation or exceeds the permitted use, you will need to obtain permission directly from the copyright holder.

Chapter 14
Reading Text in Internet Browsers

14.1 The Legibility of Serif and Sans Serif Typefaces in Internet Browsers

The research studies described thus far have been mainly concerned with material generated on local workstations using conventional word-processing software. However, many readers view material that has been saved in hypertext markup language (HTML) on remote sites on the internet to be viewed in web browsers. Even so, this is not a hard-and-fast distinction. First, word-processed documents can be uploaded to remote web sites and retrieved by other users to read on their smartphones or tablet computers as well as their own workstations. Second, how downloaded documents appear on-screen will depend on the browser settings and other software on the local device. Third, material can also be saved in HTML on local workstations and viewed through web browsers in order to mimic the retrieval of information from remote web sites. Even so, the question arises whether serif and sans serif typefaces differ in legibility when used in documents saved in HTML and viewed through web browsers. As with material that is generated on local workstations, designers and design educators tend to recommend that sans serif typefaces should be used on web sites due to the poor legibility of serif typefaces on low-resolution monitors or with small type sizes (Davidow, 2002). However, others maintain that a preference for sans serif typefaces for websites simply reflects readers' greater familiarity with sans serif typefaces when accessing sources of information on the internet (Redich, 2012, p. 62).

Gosse (1999) asked 200 participants to read stories selected from the websites of real newspapers published outside the immediate locality. Eight stories were edited to a standard length of 325 words and presented on a laptop computer with a liquid crystal display (LCD) screen using a web browser as if they were being viewed on the World Wide Web. Four stories were presented in different serif typefaces (Courier, New Century Schoolbook, Palatino, and Times), and four were presented in different sans serif typefaces (Avant Garde, Hallmarke Light, Helvetica, and Quick Type). The stories were assigned at random to four pairs, each containing a story in serif typeface and a story in sans serif typeface. Participants were timed while they read the two stories in a pair, and they were then asked a number of questions, including their preference between the two typefaces they had seen (pp. 67–75).

There was no significant difference in either reading time or preference: serif passages were read in an average of 92.0 s, and sans serif passages were read in an average of 95.8 s; 102 of the participants preferred the serif typefaces, and 98 preferred the sans serif typefaces (p. 81). Unfortunately, there were two problems with this study. First, the opening page on the web browser listed short titles of the eight stories, each presented in the appropriate typeface, and the participants were free to choose which of the stories they read first (p. 91). This then determined which story they read second and which typefaces the participant received. This contradicts the author's assertion that the order of presentation of the different serif typefaces was "systematically randomized" (p. 69). Second, in the data analyses, the contrast between serif and sans serif typefaces was incorrectly treated as a between-subjects variable and not as a within-subjects variable (pp. 84–85), and this would have reduced the analyses' statistical power.

Grant and Branch (2000) asked 21 undergraduate student teachers to participate in an online experiment using a specially constructed website. The stimuli were two passages of 164 words taken from the Graduate Record Examination and two test questions presented on the same screen as the passages. Students who used a Windows platform were shown stimuli in Times New Roman and Arial; those who used a Macintosh platform were shown stimuli in Times and Helvetica. Students who accessed the website alternately received a serif typeface first or a sans serif typeface first; in both cases, all the stimuli were shown as 11-point black text on a white background. The system recorded the reading time for each passage, and the students were given feedback on their answers to the questions, but these were not recorded. The reading data were converted to words per minute and showed that the serif typefaces were read significantly more quickly than sans serif typefaces, but there was no significant practice effect between the two passages. Grant and Branch acknowledged that the number of participants in their study was relatively small and that they had compared just two typefaces in each participant. They also had no control over the platforms used by the participants.

14.2 The Research of Bernard and Colleagues

Michael Bernard and his colleagues carried out a series of experiments to evaluate the legibility of different typefaces when reading online material. They identified passages of text (typically around 1,000 words for young adult readers) and in each case replaced 15 randomly selected words with substitutes. The latter rhymed with the original words but were semantically inappropriate to the context. The passages were saved in HTML and were viewed on a high-resolution LCD monitor using a browser as if they had been retrieved from the internet. The participants were asked to read each passage silently but to say the inappropriate words aloud. They were

14.2 The Research of Bernard and Colleagues

also asked to rate the different passages on several characteristics and to rank their overall preference among the different typefaces.

This research was initially published in *Usability News*, a biannual newsletter that was produced by the Software Usability Research Laboratory at Wichita State University. Dyson (2005) argued that these reports could not be relied upon because they had not been peer reviewed. In fact, some of the reports were also published as articles in academic journals or conference proceedings, where they will certainly have been exposed to independent peer review. In such cases, I will cite both versions of these reports so that readers can make meaningful comparisons for themselves.

Bernard and Mills (2000; Bernard et al., 2003) compared the presentation of material in either a serif typeface, Times New Roman, or a sans serif typeface, Arial, in either 10-point type or a 12-point type, and in either a aliased form or an anti-aliased form. There was no significant variation either in the participants' detection of substitutes or in the time they had taken to read the different passages. However, the fastest reading time was obtained when the passages were presented in 12-point aliased Times New Roman, and the slowest time was obtained when they were presented in 10-point anti-aliased Arial. There was no significant difference in perceived legibility between the passages presented in Arial and those presented in Times New Roman, but the passages presented in a 12-point Arial typeface (whether aliased or anti-aliased) were rated as sharper than the passages presented in 10-point anti-aliased Times New Roman. The passages presented in 12-point Arial (whether aliased or anti-aliased) or in 12-point aliased Times New Roman were the most preferred.

Bernard, Mills, Peterson, and Storrer (2001e) compared five serif typefaces (Century Schoolbook, Courier New, Georgia, Goudy Old Style, and Times New Roman), five sans serif typefaces (Agency FB, Arial, Comic Sans MS, Tahoma, and Verdana), and two ornate typefaces (Bradley Hand ITC and Monotype Corsiva). The typefaces were matched in terms of their body height and were mainly 12-point. (Whether they were aliased or anti-aliased was not specified.) There was significant variation in the time taken to read the different passages, with Corsiva yielding the fastest reading time and Tahoma the slowest. However, there was no overall difference in reading time between the serif typefaces and the sans serif typefaces. Moreover, when Bernard et al. constructed a measure of "reading efficiency" by dividing the percentage of substitutions that were detected by the overall reading time, there was no significant variation in this measure among the 12 typefaces. This suggests that any variation in reading time represented a trade-off between speed and accuracy rather than any genuine differences in reading efficiency. The participants' perceptions showed significant variation, but again there was no overall difference between the serif and sans serif typefaces. Finally, Arial, Comic Sans, Tahoma, Verdana, Courier New, Georgia, and Century Schoolbook were ranked higher than other typefaces in terms of the participants' overall preference.

Bernard, Lida, Riley Hackler, and Janzen (2002a) compared eight different typefaces. Of the four serif typefaces, two, Courier New and Times New Roman, had originally been designed for print applications; one, Century Schoolbook, had been designed for educational materials; and one, Georgia, had been designed to be displayed on computer screens. Of the four sans serif typefaces, two, Arial and Comic Sans MS, had originally been designed for print applications, and two, Tahoma and Verdana, had been designed to be displayed on computer screens. Different groups of participants were presented with passages in 10-point, 12-point, or 14-point type.

Bernard et al. found that passages presented in Times New Roman or Arial were read significantly more quickly than those presented in Courier New, Georgia, or Century Schoolbook. Again, however, there was no overall difference in reading time between the serif and sans serif typefaces. The passages presented in 12-point type were read significantly more quickly than those presented in 10-point type. The researchers noted that passages which were read more quickly tended to be read less accurately: in other words, there was a speed–accuracy trade-off. This time, they calculated a measure of reading efficiency by dividing the overall reading time by the percentage of substitutions detected (in other words, the reciprocal of their previous measure of reading efficiency). On this measure, there was no significant variation in this measure among the eight typefaces. The participants' perceptions again showed significant variation, but there was no systematic difference between either the ratings or the rankings of the serif typefaces and the sans serif typefaces.

Bernard and colleagues carried out further experiments with both older and younger participants. Bernard, Liao, and Mills (2001b, c) tested 27 adults aged between 62 and 83. Passages of around 700 words were presented in two serif typefaces, Times New Roman and Georgia, or two sans serif typefaces, Arial and Verdana, in either 12-point or 14-point type. In each passage, ten randomly selected words had been replaced by substitutes that rhymed with the original words but were semantically inappropriate to the context. The participants were also asked to rate the different passages with regard to the perceived legibility of the typeface and to rank their overall preference of the typefaces.

The passages presented in 12-point serif typefaces were read significantly less quickly than the passages presented in either 14-point serif typefaces or 14-point sans serif typefaces. When Bernard et al. constructed a measure of reading efficiency by dividing the percentage of substitutions detected by the overall reading time, the 14-point passages yielded higher scores than the 12-point passages, but there was no significant variation in reading efficiency across the four typefaces. Similarly, the participants rated the 14-point passages as being more legible than the 12-point passages, but there was no significant variation in their ratings of the four typefaces. Finally, the 14-point sans serif typefaces were ranked higher than the 12-point sans serif typefaces and all of the serif typefaces in terms of the participants' overall preference. No significant differences were found on any measure between the typefaces designed for printing on paper (Times New Roman and Arial) and those designed for screen display (Georgia and Verdana).

Bernard, Liao, Chaparro, and Chaparro (2001a) repeated this experiment with a new sample of 26 older adults in order to focus on their perceptions of different typefaces. After reading each passage, they rated it on 7-point scales in terms of its legibility, how easy it was to read, its sharpness and crispness, its attractiveness, and its personality. Finally, they ranked their overall preference of the typefaces. The 14-point passages obtained higher scores than the 12-point passages, although this was mainly true for men, not for women. No significant differences were found among the four typefaces on any of the aspects of the participants' perceptions. Overall, the participants ranked the sans serif typefaces higher than the serif typefaces, but there was no systematic difference between the ranks of the typefaces designed for printing on paper and the ranks of those designed for screen display.

Bernard, Mills, Frank, and McKown (2001d; Bernard, Chaparro, Mills, & Halcomb, 2002b) tested 27 children aged between 9 and 11. Children's short stories of about 580 words were presented in two serif typefaces, Times New Roman and Courier New, or two sans serif typefaces, Arial and Comic Sans MS, in either 12-point or 14-point type. In each story, 15 randomly selected words had been replaced by substitutes that rhymed with the original words but were semantically inappropriate to the context. The participants were also required to rate the different stories with regard to how easy they were to read, whether they enabled them to read faster, the attractiveness of the typeface, and whether they would like their schoolbooks to use the typeface. Finally, they ranked their overall preference of the eight typefaces.

There were no significant differences among the four typefaces and the two type sizes in terms of either the detection of substitutes or the speed of reading. Bernard et al. computed a measure of reading efficiency by dividing the reading time by the percentage of substitutes detected (so that lower scores implied higher efficiency). The only significant difference was that reading efficiency was less on stories presented in Courier New than on stories presented in the other typefaces. The stories presented in 14-point type were rated as significantly better than those presented in 12-point type in terms of their ease of reading, reading more quickly, their attractiveness, and their use in schoolbooks. Stories presented in Times New Roman were rated as being less easy to read than those presented in Arial or Comic Sans; stories presented in Times New Roman were rated as being less attractive than those presented in Comic Sans; and those presented in Times New Roman or Courier New were rated as less desirable for use in schoolbooks. Among the 14-point typefaces, Arial and Comic Sans were ranked higher in overall preference than Courier New or Times New Roman; among the 12-point typefaces, Comic Sans was ranked higher in overall preference than the other typefaces.

14.3 Subsequent Research

Myung (2003) presented 12 Korean students with newspaper stories of between 453 and 532 words using the internet browser installed on a personal computer. The stories were shown in three typefaces: one serif typeface (Batang) and two sans serif typefaces (Dodum and Gulim). The participants were asked to read the stories to themselves and then to rate their typographical appearance on a 7-point scale. Myung calculated a measure of reading speed by dividing the total number of characters in each story by the time taken to read it. There was no significant variation among the three typefaces in terms of the participants' reading speed. However, the application of conjoint analysis to the participants' preference ratings showed that the stories that had been printed in Dodum and Gulim were preferred to those printed in Batang.

Ling and van Schaik (2006) carried out two experiments to examine the influence of typeface and line length on students' use of web pages. In both cases, the web pages were presented in either a serif typeface (12-point Times New Roman) or a sans serif typeface (10-point Arial) and in four different line lengths. In their first experiment, 72 participants had to say whether or not a mock web page presented in a browser contained a specified hyperlink. There were no significant differences between the students who saw web pages in the serif typeface and those who saw web pages in the sans serif typeface in either the accuracy or the response time for hits or in either the accuracy or the response time for correct rejections.

In their second experiment, 99 participants had to answer questions based upon the information contained in five mock web sites, each consisting of 30 pages, on various topics. The proportion of correct answers approached 100%. There were no significant differences between the students who saw web sites in the serif typeface and those who saw web sites in the sans serif typeface in either the time taken to carry out their task or the number of web pages that they visited to find the correct answers. Ling and van Schaik concluded that there was no difference between the serif typeface and the sans serif typeface either in visual search or in information retrieval.

After both experiments, the participants were asked to express a preference between the two typefaces and to rate their aesthetic value on a 10-point scale. In the first experiment, regardless of which typeface they had seen, the participants tended to prefer Arial rather than Times New Roman and to rate Arial more highly than Times New Roman in terms of aesthetic value, although the latter difference was small in magnitude and unlikely to be of any practical importance. In the second experiment, there were no significant differences in either the participants' preference or in their ratings of aesthetic value.

Chernecky et al. (2006) recruited 22 cancer patients. The patients were assigned to workstations in groups of two or three but recorded their individual responses on a prepared form. The stimuli were presented using an internet browser, but the computers and monitors used were not specified. In one section of the test, the patients were presented with examples of text in varying sizes and in different typefaces with different backgrounds, two at a time. In each case, they indicated which of the two displays that they preferred. The strongest preference was for the serif typeface Times New Roman in a ten-point font and in blue lettering on either a tan or white background. The next strongest preference was for the sans serif typeface Arial in a nine-point font and black lettering on a tan background. However, the sans serif typeface Verdana was not preferred. Chernecky et al. ascribed the preference for the serif typeface to the fact that it was widely used in books, newspapers, and magazines. They did not carry out any kind of statistical analysis of their results, and they did not include any comparison group, and so it is unclear whether their findings were peculiar to cancer patients or would generalise to other kinds of participant.

In a study mentioned in Sect. 12.5, Shaikh et al. (2006) obtained participants' ratings of the appropriateness of different typefaces for various online purposes.

14.3 Subsequent Research

Fox et al. (2007) selected three of these typefaces judged to be of high, medium, or low appropriateness for each of three purposes: a business document, an e-mail message, and a narrative for young people. A total of 120 participants were presented with an example of each document (a bank letter, an e-mail invitation to a company picnic, and an explanation of how fireworks work) in one of the three typefaces. Each was presented as an HTML web page and was viewed using a web browser. The participants were asked to rate the "personality" of the document using 15 bipolar scales from Shaikh et al.'s (2006) study and to rate its "ethos" (their perceptions of the author and the intended readership) using five scales.

The choice of typeface had no significant effect on the participants' perceptions of the business document, except that its author was viewed as less mature if the least appropriate typeface was used. If the least appropriate typeface was used for the e-mail message, it was viewed as less stable, less practical, more rebellious, more youthful, and more feminine; its author was also viewed as less believable, less professional, less trustworthy, and less mature. The choice of typeface had no significant effect on perceptions of the narrative for young people, except that it was viewed as more youthful and more casual if the most appropriate typeface was used. Fox et al. concluded that in general on-screen documents were more likely to be perceived in a negative manner if they were presented using a less appropriate typeface.

Beymer et al. (2008) carried out a similar study. They presented 82 employees of a computer company with one-page stories taken from a science news website. They were presented on a computer screen as a series of web pages in a 12-point anti-aliased typeface: half of the participants saw the stories in the serif typeface Georgia, and half saw them in the sans serif typeface Helvetica. Their eye movements were monitored while they read the stories, and a multiple-choice test was administered after each story to check their retention. There was no significant difference between the two subgroups in their reading speed, in a variety of statistics relating to their eye movements, or in their retention. Beymer et al. noted that around half of their participants reported having a first language other than English. Having found significant differences in eye movements related to the participants' age, they focused on those aged 30–39. The participants for whom English was the first language produced shorter fixations and longer eye movements than those who had some other first language. However, neither group showed any differences between the two typefaces.

Ali et al. (2013) compared the legibility of serif and sans serif typefaces in 48 Malaysian students reading texts of moderately high difficulty containing 140 words in the Malay language. These were presented in a web browser on an LCD monitor in a 12-point typeface. For the first 24 participants, the texts were presented in two typefaces designed for screen presentation: the serif typeface Georgia and the sans serif typeface Verdana. For the second 24 participants, the texts were presented in two typefaces designed for printed media: the serif typeface Times New Roman and the sans serif typeface Arial. The participants were required to read the texts aloud as quickly and accurately as possible, and their performance was monitored by two research assistants. Their reading speed and their accuracy were both mapped onto scales from 1 to 5, and the results were added together to yield an overall score.

There was no sign of any difference in performance either between Georgia and Verdana or between Times New Roman and Arial. Two problems with this study are that all the students read the texts in the same sequence and that all saw the two typefaces in the same sequence; hence, there was no control for transfer effects (such as the positive effect of practice or the negative effect of fatigue). The researchers also carried out independent sample tests when the observations had been obtained by repeated testing of the same participants, and this would once again have reduced the analyses' statistical power.

Mátrai and Kosztyán (2014) devised web pages containing verbal comprehension tasks and compared text presented in the serif typeface Times New Roman and text presented in the sans serif typeface Arial. In addition, they manipulated the size of the text (three sizes for each typeface), the line length and spacing, the colour of the background, and the alignment of the text, which yielded a total of 144 conditions. Each of 125 university students was asked to solve the tasks on a sample of 40 web pages. (The computers and the monitors were not specified.) A regression analysis found no significant difference between the two typefaces in terms of the response latencies. Mátrai and Kosztyán stated that there was a significant difference in terms of the proportions of correct responses, but they did not provide any further information. In fact, the overall difference was relatively slight (Times New Roman, 84.0%; Arial, 83.3%) and unlikely to be of any practical importance. Finally, the students were asked to express a preference among the different displays, but there was no significant difference in their preference between the two typefaces. A fundamental problem with this study is that Mátrai and Kosztyán assigned different verbal comprehension tasks to different conditions, but they failed to evaluate whether the tasks were of equal difficulty. Consequently, even the small difference that they found between the two typefaces in terms of the proportions of correct responses might have been due to differences in the difficulty of the relevant tasks rather than to differences in the legibility of the typefaces.

Beyon and Cox-Boyd (2020) carried out a follow-up to the study mentioned in Sect. 6.4 by Gasser et al. (2005), who varied the typeface used when participants were reading text from paper. Beyon and Cox-Boyd used a text concerning spinal health. They presented this text in four different typefaces: a monospaced slab serif typeface (Courier New), a monospaced sans serif typeface (Lucida Console), a proportionally spaced serif typeface (Palatino Linotype), and a proportionally spaced sans serif typeface (Arial). Independent of this, they presented the text in black, blue, or red, yielding 12 different conditions. They recruited volunteers from an online website (Amazon Mechanical Turk), who were asked to carry out the task as an online survey. They were asked to read the text, complete a questionnaire about their attitudes to spinal health as a distractor task, and then answer six questions to test their retention of the key information contained in the original document. The participants were randomly assigned to one of the 12 presentation conditions. Beyon and Cox-Boyd found no significant differences in performance among the four typefaces or the three type colours. Beyon and Cox-Boyd acknowledged that they had no control over the devices or platforms which the participants had used to carry out the task.

14.4 Conclusions

This chapter discussed whether serif and sans serif typefaces differ in their legibility when the material is saved in HTML and viewed on-screen through web browsers. This includes material saved in local workstations as well as material retrieved from the internet. In addition to a variety of individual studies, the chapter described a research programme that was carried out by Bernard and colleagues at Wichita State University. Further research has been carried out into the use of different typefaces for various online purposes. When reading material in internet browsers, by far the most common finding is that there is no significant difference between serif typefaces and sans serif typefaces in terms of the users' reading comprehension, reading speed, or reading accuracy. There is also no consistent evidence that readers have a preference between serif typefaces and sans serif typefaces when reading material in internet browsers. Once again, both serif and sans serif typefaces are regarded as being broadly appropriate for internet sites, whereas display and cursive typefaces are regarded as being generally inappropriate for serious use.

Open Access This chapter is licensed under the terms of the Creative Commons Attribution 4.0 International License (http://creativecommons.org/licenses/by/4.0/), which permits use, sharing, adaptation, distribution and reproduction in any medium or format, as long as you give appropriate credit to the original author(s) and the source, provide a link to the Creative Commons license and indicate if changes were made.

The images or other third party material in this chapter are included in the chapter's Creative Commons license, unless indicated otherwise in a credit line to the material. If material is not included in the chapter's Creative Commons license and your intended use is not permitted by statutory regulation or exceeds the permitted use, you will need to obtain permission directly from the copyright holder.

Chapter 15
General Conclusions to Part II

15.1 Key Findings from Part II

Once again, as was mentioned in Sect. 1.4, Part II of this book has reviewed diverse studies using diverse methods of data collection, and this precludes any formal meta-analysis to integrate the findings. One must instead focus on the most common finding—the *modal* finding—regarding the legibility of serif and sans serif typefaces: superiority of serif typefaces; superiority of sans serif typefaces; or no difference. Part II has been concerned with the question of whether there are differences in the legibility of serif and sans serif typefaces when they are used to produce material to be read on computer monitors or other screens.

Studies of the legibility of letters and words have proved inconclusive (Sect. 11.1). The studies that employed authentic typefaces showed at most that *some* sans serif typefaces (Consolas, Letter Gothic, and Verdana) are more legible than *some* serif typefaces (Times New Roman and the slab serif typeface Courier). In addition, researchers who employed artificial typefaces confounded the presence or absence of serifs with variations in the width of the letters and variations in the spacing among successive letters. Early studies using the tachistoscopic presentation of *printed* letters and words showed that errors in their identification were often the result of confusions among visually similar letters (see Sect. 4.2), and this idea has been confirmed using *screen-based* presentation (Sect. 11.2). However, such confusions seem to be due to the design of individual characters in specific typefaces rather than to the presence or absence of serifs. Indeed, visual confusions are not more likely with sans serif typefaces than with serif typefaces, which contradicts the old hypothesis that serifs make letters easier to discriminate and identify (Legros, 1922, p. 11).

Early research using print-based presentation suggested that visual confusions were much less important when reading connected sentences (Vernon, 1929), but this notion does not seem to have been tested using presentation on computer screens. Five studies compared the *speed* with which participants read sentences displayed in serif and sans serif typefaces (Sect. 12.1): one found that serif typefaces were read more quickly, two found that sans serif typefaces were read more quickly, while two found no significant difference. Three of these studies also compared the participants' *accuracy*: one found that sans serif typefaces were read more accurately, while two found no significant difference. Research into readers' eye movements has not yielded consistent findings with regard to the presence or absence of serifs.

Five studies compared readers' comprehension of meaningful text when presented on computer screens in serif and sans serif typefaces (Sect. 12.2). Four of the studies found no significant difference in the time taken to read the material in question. The fifth (Hojjati & Muniandy, 2014) found that material displayed in a sans serif typeface was read more quickly, but this study suffered from serious methodological problems. Three of the studies reported measures of comprehension: one found an advantage for a serif typeface, one found an advantage for a sans serif typeface, and the third found no significant difference.

The rapid serial visual presentation (RSVP) procedure constitutes another means of presenting meaningful material, and five studies have compared readers' performance with serif and sans serif typefaces in terms of their comprehension, speed, or accuracy (Sect. 12.3). All five studies found no significant difference between serif and sans serif typefaces, except that serif styles were less legible with very small type or under conditions of low luminance (when, of course, serifs are likely to have been faint or even completely invisible). Two Korean studies compared the legibility of serif typefaces and sans serif typefaces when viewed on the screen of a personal digital assistant or a smartphone, but neither study found any significant difference in the participants' reading performance (Sect. 12.4).

Differences in the legibility of serif and sans serif typefaces are not apparent even in readers whose visual systems are challenged as the result of disablement. Different studies have examined a heterogeneous sample of patients with visual impairment (Sect. 13.1), a large sample of readers with dyslexia (Sect. 13.2), and people with and without age-related macular degeneration (Sect. 13.3). In each case, there was no difference in the participants' reading speed between serif typefaces and sans serif typefaces. Any differences in the participants' preferences or eye movements could be attributed to the researchers' failure to control the width or inter-letter spacing of different typefaces.

Finally, when reading material in internet browsers, by far the most common finding is that there is no significant difference between serif typefaces and sans serif typefaces in terms of the users' reading comprehension, reading speed, or reading accuracy (Chap. 14). This applies regardless of whether or not researchers have used measures of reading "efficiency" to control for the possibility that readers employ some kind of trade-off between their speed and their accuracy. There is also no consistent evidence that readers have a preference between serif typefaces and sans serif typefaces when reading material in internet browsers. Once again, both serif and sans serif typefaces are regarded as being broadly appropriate for internet sites, whereas display and cursive typefaces are regarded as being generally inappropriate for serious use.

15.2 Preferences and Connotations

Some studies, though not all, have found that readers express a preference for sans serif typefaces when reading on computer screens, but in general both serif and sans serif typefaces are regarded by users as appropriate for online purposes (Sect. 12.5).

There is some evidence that serif and sans serif typefaces differ in their connotations or "personality". This seems to reflect variations in reader' expectations, which in turn depend on their prior experience and familiarity with different typefaces. When reading on computer screens, these differences are small in magnitude and are often not statistically significant. One study (Kaspar et al., 2015) found that scientific abstracts were rated more positively in a serif typeface than in a sans serif typeface when artificial typefaces were used, but the reverse was true when authentic typefaces were used. This suggests that other features can override any effect of the presence or absence of serifs.

One limitation of the latter study is that the ratings were provided by students and not by experienced teachers or researchers. Nevertheless, it raises the possibility that evaluations of academic writing may be influenced by readers' preferences and the connotations of serif and sans serif typefaces. Section 9.2 discussed the possible implications of this notion in the context of reading from paper, but similar points can be made about reading on-screen:

- In the context of academic publication, authors will want to be assured that their work is evaluated in terms of its content rather than in terms of its typographical appearance. The findings of Kaspar et al. (2015) indicate that on-screen evaluations of scientific abstracts can be influenced by the typeface in which they are presented. It is reasonable to assume that the same applies to on-screen evaluations of entire articles or books (although there seems to be no empirical evidence on this matter). One solution to this problem is for academic journals and publishers to require that manuscripts should be submitted for publication in a standard typeface so that they can be compared on a like-for-like basis.
- In the context of academic assessment, it is possible that teachers and other assessors will give more positive on-screen evaluations of students' online assignments if the teachers and their students share the same typographical preferences than if they differ in those preferences (although once again there seems to be no empirical evidence on this matter). It would be both useful and fairer in the interests of ensuring valid assessment if teachers responsible for particular course units (and, ideally, for entire degree programmes) could agree on their typographical preferences and make these known to their students.

Of course, in both contexts these variations in readers' expectations and preferences might depend on their prior experience and familiarity with different typefaces rather than on any intrinsic properties of the typefaces themselves (and there *is* empirical evidence on this point in the case of reading from paper, if not in the case of reading from screens: see Sect. 9.2). Even so, there is a need for research on the extent to which reviewers' on-screen evaluations of academic manuscripts and teachers' on-screen evaluations of their students' assignments are affected by the reviewers' and teachers' preferences and expectations.

15.3 Implications for Previous Assumptions

Where does this leave the recommendations of designers and design educators who have traditionally advocated the use of sans serif typefaces for material presented on-screen? Poncelet and Proctor (1993, p. 101) simply stated without qualification or supporting evidence: "Often san-serif fonts work better on the computer screen than serif fonts." This assertion is clearly not supported by the research evidence reviewed in Part II. Similarly, Schriver (1997) claimed without qualification or supporting evidence that sans serif was "the preferred style of type for online because of its simple, highly legible, modern appearance" (p. 508). Her claims about the appearance of sans serif typefaces are partially supported by the results of research on the connotations of different typefaces, but they do not appear to translate into objective differences in their legibility.

Universal design is an approach to the development of educational websites that aims to ensure accessibility for all students instead of implementing ad hoc adjustments for those with particular disabilities. Proponents of universal design have advocated: "Use Sans Serif fonts for text. Letters with serifs are difficult to read on-screen and can create visual fatigue when large amounts of text are included on web sites" ("Universal Design", 1999, p. 6). In general, there is no support for the idea that serif typefaces are harder to read on-screen than sans serif typefaces. One might expect that such effects would be more evident in readers with visual impairment under the high demands of the RSVP procedure, yet two studies failed to find any significant difference in performance between serif typefaces and sans serif typefaces in such readers. The quoted text refers specifically to material included on web sites, yet there is no evidence for differences between serif typefaces and sans serif typefaces when reading from internet browsers.

In fact, the notion of universal design has had its critics. For instance, Raymaker et al. (2019, p. 148) argued that it was

> ultimately impractical due to the fact that access needs can conflict with each other. For example, some guidelines intended for people with intellectual disability . . . recommend simplifying vocabulary, which—if implemented without retaining the precision afforded by more complex wording—can make language pragmatics more difficult for autistic users to understand. . . . Likewise, high-contrast color schemes suitable for people with low vision may be painful or unreadable to autistic users with hypersensitive vision

Even so, Raymaker et al. concurred that websites designed for people with autism should "use a plain accessible sans-serif font (e.g., Arial) for ease of readability" (p. 147). Once again, they provided no empirical evidence to support this recommendation, and so we are back in the world of "everybody knows".

15.4 Conclusions

This chapter concludes Part II by summarising and discussing the key findings. Studies of the legibility of letters and words when presented on computer monitors or other screens have failed to yield a consistent pattern of results, as have studies of the legibility of connected sentences, even when using the RSVP procedure. With regard to readers' comprehension of meaningful text when presented on computer screens, the modal finding is that of no significant difference in reading speed between serif and sans serif typefaces. The studies that reported measures of accuracy in this situation failed to show a consistent pattern of results. Some studies have found that readers express a preference for sans serif typefaces, and there is some evidence that serif and sans serif typefaces differ in their connotations; however, these findings can be attributed to readers' previous experience of reading text on-screen. Previously stated assumptions about the legibility of serif and sans serif typefaces when used to present material on computer screens are not supported by empirical research findings. In short, despite various assertions and recommendations, the conclusion of Part II is that there is no difference in the legibility of serif typefaces and sans serif typefaces when they are used to produce material that is presented on computer monitors or other screens.

Open Access This chapter is licensed under the terms of the Creative Commons Attribution 4.0 International License (http://creativecommons.org/licenses/by/4.0/), which permits use, sharing, adaptation, distribution and reproduction in any medium or format, as long as you give appropriate credit to the original author(s) and the source, provide a link to the Creative Commons license and indicate if changes were made.

The images or other third party material in this chapter are included in the chapter's Creative Commons license, unless indicated otherwise in a credit line to the material. If material is not included in the chapter's Creative Commons license and your intended use is not permitted by statutory regulation or exceeds the permitted use, you will need to obtain permission directly from the copyright holder.

Chapter 16
Coda: Lessons Learned

Both serif typefaces and sans serif typefaces have a long history, going back to inscriptions in Ancient Rome and probably earlier (see Sects. 1.2 and 1.3). Historically, serif typefaces were used in printing earlier than sans serif typefaces, and they have usually been preferred to sans serif typefaces for use in formal documents. Section 1.2 described a number of theories aimed at explaining why serifs should have survived in modern typography and indeed at explaining why serif typefaces should be more legible than sans serif typefaces when reading from paper: (a) that serifs constitute additional visual cues to support the reader's gaze; (b) that they overcome the harmful effects of irradiation; and (c) that they facilitate the operation of line detectors in the human visual system. However, none of these explanations has proved especially convincing in the light of subsequent arguments and evidence. In the case of reading from screens, it was argued that serifs and other details might be lost when reading from low-resolution monitors, but this is not a plausible argument now that high-resolution monitors are widely available.

A different approach is to claim that serif and sans serif typefaces do not differ in their legibility because of inherent properties of serifs themselves, but that the presence or absence of serifs serves as a proxy for some other property of typefaces. This approach was adopted by Wilkins et al. (2020; see Sect. 11.2), who suggested that serifs tended to accentuate the spatial periodicity of letter strokes (i.e., the vertical "stripiness" of words), as measured by the first peak in a word's horizontal autocorrelation. Wilkins et al. showed that words of low vertical stripiness are read more quickly than words of high vertical stripiness, and that serif typefaces tend to have higher stripiness than sans serif typefaces, but they did not show that this leads to variations in how quickly words in different typefaces are read. For the moment, the theoretical and practical implications of vertical stripiness must remain unclear and contentious. More important, the arguments put forward by Wilkins et al. entail that serif typefaces should be less legible than sans serif typefaces *both* when reading from paper *and* when reading from screens, which is not position that has been adopted to date by any other researchers.

In contrast, there is a good deal of evidence about the negative effects of *horizontal* stripiness—the extent to which successive lines of text tend to resemble horizontal stripes. As mentioned in Sect. 4.1, horizontal stripes are known to induce eye strain, visual illusions, headache, and even seizures. Wilkins et al. (1984) used the Michelson contrast to measure the horizontal "stripiness" of visual displays: this is defined as the difference in the luminances of the light and dark sections of a pattern divided by the sum of their luminances. In the case of horizontal stripes, Wilkins et al. found a clear relationship between this measure and the probability of illusions and seizures. Wilkins and Nimmo-Smith (1987) applied the measure to lines of printed text and found that it was inversely related to readers' ratings of the clarity and comfort of the material. More recently, Wilkins et al. (2020) applied Fourier analysis to text displayed on an LCD screen and obtained similar results. In both studies, the clarity and comfort of the text appeared to be improved by increasing the spacing between the lines of text relative to the x-height of the typeface. However, Wilkins and Nimmo-Smith (1987) had compared the mean Michelson contrast of samples of 24 serif typefaces and seven sans serif typefaces: they found a highly significant variation across the 31 typefaces, but they did not find any systematic difference between the serif typefaces and the sans serif typefaces. While clarity and comfort may be very important characteristics of text presented on paper or on screens, there seems to be no difference between serif and sans serif typefaces in these characteristics.

In fact, the copious evidence that has been reviewed in this monograph leads to the conclusion that there is no difference in the legibility of serif and sans serif typefaces *either* when reading from paper *or* when reading from screens. This contradicts various assertions made over the last 100 years by typographers, designers, and other authority figures. What this means is that assertions to the effect that "everyone knows" that such-and-such should not be accepted on the basis of the authority or status of the people or organisations making them but should be regarded merely as conjectures that might well be subject to refutation through carefully designed empirical research (cf. Popper, 1959, 1962.) Of the large number of studies that I have reviewed in this monograph, some have been more carefully designed than others, and I have identified at least some of the design flaws that are apparent in previous research. Nevertheless, the overwhelming thrust of the available evidence is that there is no difference in the legibility of serif typefaces and sans serif typefaces *either* when reading from paper *or* when reading from screens. Typographers and software designers should feel able to make full use of both serif typefaces and sans serif typefaces, even if legibility is a key criterion in their choice.

Open Access This chapter is licensed under the terms of the Creative Commons Attribution 4.0 International License (http://creativecommons.org/licenses/by/4.0/), which permits use, sharing, adaptation, distribution and reproduction in any medium or format, as long as you give appropriate credit to the original author(s) and the source, provide a link to the Creative Commons license and indicate if changes were made.

The images or other third party material in this chapter are included in the chapter's Creative Commons license, unless indicated otherwise in a credit line to the material. If material is not included in the chapter's Creative Commons license and your intended use is not permitted by statutory regulation or exceeds the permitted use, you will need to obtain permission directly from the copyright holder.

References

Accessibility. (2020). Washington, DC: American Psychological Association. Retrieved June 30, 2021, from https://apastyle.apa.org/style-grammar-guidelines/paper-format/accessibility.

Adams, M. J. (1979). Models of word recognition. *Cognitive Psychology,11*(2), 133–176. https://doi.org/10.1016/0010-0285(79)90008-2

Adams, S., Rosemier, R., & Sleeman, P. (1965). Readable letter size and visibility for overhead projection transparencies. *AV Communication Review,13*(4), 412–417. https://doi.org/10.1007/BF02766846

Akhmadeeva, L., Tukhvatullin, I., & Veytsman, B. (2012). Do serifs help in comprehension of printed text? An experiment with Cyrillic readers. *Vision Research,65*, 21–24. https://doi.org/10.1016/j.visres.2012.05.013

Ali, A. Z. M., Wahid, R., Samsudin, K., & Idris, M. Z. (2013). Reading on the computer screen: Does font type has [*sic*] effects on web text readability? *International Education Studies,6*(3), 26–35. https://doi.org/10.5539/ies.v6n3p26

American Psychological Association. (2010). *Publication manual of the American Psychological Association* (6th ed.).

American Psychological Association. (2020). *Publication manual of the American Psychological Association* (7th ed.). https://doi.org/10.1037/0000165-000

Aquilante, K., Wyatt, H., & Yager, D. (1997). Eye movements of congenital nystagmats are slower during RSVP reading than during PAGE reading. In D. Yager (Ed.), *Non-invasive assessment of the visual system* (*Trends in Optics and Photonics*, Vol. 11, pp. 127–131). Optical Society of America.

Arditi, A. (2004). Adjustable typography: An approach to enhancing low-vision text. *Ergonomics,47*(5), 469–482. https://doi.org/10.1080/0014013031000085680

Arditi, A., & Cho, J. (2005). Serifs and font legibility. *Vision Research,45*(23), 2926–2933. https://doi.org/10.1016/j.visres.2005.06.013

Arnold, E. C. (1956). *Functional newspaper design*. Harper.

Aten, T. R., Gugerty, L., & Tyrrell, R. A. (2002). Legibility of words rendered using ClearType™. *Proceedings of the Human Factors and Ergonomics Society,46*(17), 1684–1687. https://doi.org/10.1177/154193120204601733

Bailey, I. L., & Lovie, J. E. (1976). New design principles for visual acuity letter charts. *American Journal of Optometry and Physiological Optics,53*(11), 740–745.

Balfour, S., Borthwick, S., Cubelli, R., & Della Sala, S. (2007). Mirror writing and reversing single letters in stroke patients and normal elderly. *Journal of Neurology,254*(4), 436–441. https://doi.org/10.1007/s00415-006-0384-8

Banerjee, J., & Battacharyya, M. (2011). Selection of the optimum font type and size interface for on screen continuous reading by young adults: An ergonomic approach. *Journal of Human Ergology, 40*(1–2), 47–62. https://doi.org/10.11183/jhe.40.47

Baron, N. S., Calixte, R. M., & Havewala, M. (2017). The persistence of print among university students: An exploratory study. *Telematics and Informatics,34*(5), 590–604. https://doi.org/10.1016/j.tele.2016.11.008

Bartram, D. (1982). The perception of semantic quality in type: Differences between designers and non-designers. *Information Design Journal,3*(1), 38–50. https://doi.org/10.1075/idj.3.1.04bar

Beier, S., & Dyson, M. C. (2014). The influence of serifs on "h" and "i": Useful knowledge from design-led scientific research. *Visible Language,47*(3), 74–95.

Beier, S., & Larson, K. (2010). Design improvements for frequently misrecognized letters. *Information Design Journal,18*(2), 118–137. https://doi.org/10.1075/idj.18.2.03bei

Bell, R. C., & Sullivan, J. L. F. (1981). Student preferences in typography. *Programmed Learning and Educational Technology,18*(2), 57–61. https://doi.org/10.1080/0033039810180201

Bennett, A. G. (1965). Ophthalmic test types: A review of previous work and discussions on some controversial questions. *British Journal of Physiological Optics,22*(4), 238–271.

Benton, C. L. (1979, August 5–8). *The connotative dimensions of selected display typefaces*. Paper presented at the 62nd Annual Meeting of the Association for Education in Journalism, Houston, TX (ED172262). ERIC https://files.eric.ed.gov/fulltext/ED172262.pdf.

Berliner, A. (1920). "Atmosphärenwert" von Drucktypen: Ein Beitrag zur Psychologie der Reklame. ["Atmosphere value" of typefaces: A contribution to the psychology of advertising]. *Zeitscrift für angewandte Psychologie, 17*(1–3), 165–172.

Bernard, J.-B., Aguilar, C., & Castet, E. (2016). A new font, specifically designed for peripheral vision, improves peripheral letter and word recognition, but not eye-mediated reading performance. *PLoS ONE, 11*(4), Article e0152506. https://doi.org/10.1371/journal.pone.0152506

Bernard, M., & Mills, M. M. (2000). So, what size and type of font should I used on my website? *Usability News*, Vol. 2, No. 2. Retrieved June 30, 2021, from https://web.archive.org/web/20080708185021/http://www.surl.org/usabilitynews/22/font.asp.

Bernard, M., Chaparro, B., & Russell, M. (2000). Is RSVP a solution for reading from small displays? *Usability News*, Vol. 2, No. 2. Retrieved June 30, 2021, from https://web.archive.org/web/20080820043225/http://www.surl.org/usabilitynews/22/rsvp.asp.

Bernard, M., Liao, C. H., Chaparro, B. S., & Chaparro, A. (2001a, November 18–20). *Examining perceptions of online text size and typeface legibility for older males and females*. Paper presented at the 6th Annual International Conference on Industrial Engineering: Theory, Applications, and Practice, San Francisco, CA. Retrieved June 30, 2021, from https://pdfs.semanticscholar.org/c3f3/f824e0327ea45f12f341cb033b71d7adba11.pdf?_ga=2.242662914.1690164776.1578918359-946503918.1535527115.

Bernard, M., Liao, C., & Mills, M. (2001b). Determining the best online font for older adults. *Usability News*, Vol. 3, No. 1. Retrieved June 30, 2021, from https://web.archive.org/web/20080708185645/http://www.surl.org/usabilitynews/31/fontSR.asp.

Bernard, M., Liao, C., & Mills, M. (2001c). The effects of font type and size on the legibility and reading time of online text by older adults. In *CHI'01: Human Factors in Computing Systems* (pp. 175–176). Association for Computing Machinery. https://doi.org/10.1145/634067.634173

Bernard, M., Mills, M., Frank, T., & McKown, J. (2001d). Which fonts do children prefer to read online? *Usability News*, Vol. 3, No. 1. Retrieved June 30, 2021, from https://web.archive.org/web/20080708184429/http://www.surl.org/usabilitynews/31/fontJR.asp.

Bernard, M., Mills, M., Peterson, M., & Storrer, K. (2001e). A comparison of popular online fonts: Which is best and when? *Usability News*, Vol. 3, No. 2. Retrieved June 30, 2021, from https://web.archive.org/web/20080708184837/http://www.surl.org/usabilitynews/32/font.asp.

Bernard, M., Lida, B., Riley, S., Hackler, T., & Janzen, K. (2002a). A comparison of popular online fonts: Which size and type is best? *Usability News*, Vol. 4, No. 1. Retrieved June 30, 2021, from https://web.archive.org/web/20080708185245/http://www.surl.org/usabilitynews/41/onlinetext.asp.

Bernard, M. L., Chaparro, B. S., Mills, M. M., & Halcomb, C. G. (2002b). Examining children's reading performance and preference for different computer-displayed text. *Behaviour and Information Technology,21*(2), 87–96. https://doi.org/10.1080/01449290210146737

References

Bernard, M. L., Chaparro, B. S., Mills, M. M., & Halcomb, C. G. (2003). Comparing the effects of text size and format on the readability of computer-displayed Times New Roman and Arial text. *International Journal of Human-Computer Studies,59*(6), 823–835. https://doi.org/10.1016/S1071-5819(03)00121-6

Beymer, D., Russell, D., & Orton, P. (2008). An eye tracking study of how font size and type influence online reading. In *Proceedings of the 22nd British HCI Group Annual Conference on People and Computers: Culture, Creativity, Interaction* (Vol. 2, pp. 15–18). British Computer Society. https://doi.org/10.14236/ewic/HCI2008.23

Beyon, J., & Cox-Boyd, C. (2020). The influence of font type and color on online information recall. *North American Journal of Psychology, 22*(1), 13–26. Retrieved June 30, 2021, from https://www.researchgate.net/publication/237229931_The_Influence_of_Font_Type_on_Information_Recall.

Bigelow, C. (2020a). The Font Wars, part 1. *IEEE Annals of the History of Computing,42*(1), 7–24. https://doi.org/10.1109/MAHC.2020.2971202

Bigelow, C. (2020b). The Font Wars, part 2. *IEEE Annals of the History of Computing,42*(1), 25–40. https://doi.org/10.1109/MAHC.2020.2971745

Bigelow, C. (with Seybold, J.) (1981). Technology and the aesthetics of type: Maintaining the "tradition" in the age of electronics. *Seybold Report, 10*(24), 3–16. Retrieved June 30, 2021, from https://bigelowandholmes.typepad.com/bigelow-holmes/2015/05/digital-type-archaeology-1.html.

Bigelow, C., & Holmes, K. (1986). The design of Lucida®: An integrated family of types for electronic literacy. In J. C. van Vliet (Ed.), *Text processing and document manipulation* (pp. 1–17). Cambridge University Press.

Bluhm, A. (1991). Types for easy reading. *Graphic Repro,11*(3), 29–31.

Boyarski, D., Neuwirth, C., Forlizzi, J., & Regli, S. H. (1998). A study of fonts designed for screen display. In *Proceedings of the SIGCHI Conference on Human Factors in Computing Systems* (pp. 87–94). Association for Computing Machinery. https://doi.org/10.1145/274644.27465

Bringhurst, R. (2019). *The elements of typographic style* (4th ed., Version 4.3). Hartley & Marks.

British Dyslexia Association. (2018). *Dyslexia style guide 2018: Creating dyslexia friendly content*. Retrieved June 30, 2021, from https://cdn.bdadyslexia.org.uk/documents/Advice/style-guide/Dyslexia_Style_Guide_2018-final-1.pdf?mtime=20190409173950&focal=none.

British Standards Institute. (1968). *Visual acuity test types–Part 1: Specification for test charts for clinically determining distance visual acuity* (BS 4274-1:1968). British Standards Institute.

Brockington, G. (2012). Rhyming pictures: Walter Crane and the universal language of art. *Word and Image,28*(4), 359–373. https://doi.org/10.1080/02666286.2012.740185

Brookshire, C. E., Wilson, J. P., Nadeau, S. E., Gonzalez Rothi, L. J., & Kendall, D. L. (2014). Frequency, nature, and predictors of alexia in a convenience sample of individuals with chronic aphasia. *Aphasiology,28*(4), 1464–1480. https://doi.org/10.1080/02687038.2014.945389

Brown, E., Foulsham, T., Lee, C.-S., & Wilkins, A. (2020). Research note: Visibility of temporal light artefact from flicker at 11 kHz. *Lighting Research and Technology,52*(3), 371–376. https://doi.org/10.1177/1477153519852391

Brumberger, E. (2004). The rhetoric of typography: Effects on reading time, reading comprehension, and perceptions of ethos. *Technical Communication,51*(1), 13–24.

Brumberger, E. R. (2003a). The rhetoric of typography: The awareness and impact of typeface appropriateness. *Technical Communication,50*(2), 224–231.

Brumberger, E. R. (2003b). The rhetoric of typography: The persona of typeface and text. *Technical Communication,50*(2), 206–223.

Burt, C. (1959). *A psychological study of typography*. Cambridge University Press.

Burt, C. (1960). The typography of children's books: A record of research in the U.K. In G. Z. F. Bereday & J. A. Lauwerys (Eds.), *The year book of education 1960: Communication media and the school* (pp. 242–256). Evans Brothers.

Burt, C., Cooper, W. F., & Martin, J. L. (1955). A psychological study of typography. *British Journal of Statistical Psychology,8*(1), 29–56. https://doi.org/10.1111/j.2044-8317.1955.tb00160.x

Campbell, K. A., Cutler, F., McDonald, R., Putt, C., Rewak, M., Strong, G., & Whitton, H. (2006, August 10–13). *What typeface is preferred by people with macular degeneration?* [Poster presentation]. American Psychological Association 114th Annual Meeting, New Orleans, LA. https://doi.org/10.1037/e513242007-001

Caplan, P. J. (1979). Beyond the box score: A boundary condition for sex differences in aggression and achievement striving. In B. A. Maher (Ed.), *Progress in experimental personality research* (Vol. 9, pp. 41–87). Academic Press.

Caplan, P. J., Crawford, M., Hyde, J. S., & Richardson, J. T. E. (1997). *Gender differences in human cognition.* Oxford University Press.

Catich, E. M. (1991). *The origin of the serif: Brush writing & Roman letters* (2nd ed.; M. W. Gilroy, Ed.). St. Ambrose University, Catich Gallery.

Cattell, J. M. (1885). The inertia of the eye and brain. *Brain,8*(3), 295–312.

Chaparro, B. S., Shaikh, A. D., & Chaparro, A. (2006a). Examining the legibility of two new ClearType fonts. *Usability News*, Vol. 8, No. 1. Retrieved June 30, 2021, from https://web.archive.org/web/20080725064707/http://www.surl.org/usabilitynews/81/legibility.asp.

Chaparro, B. S., Shaikh, A. D., & Chaparro, A. (2006b). The legibility of ClearType fonts. *Proceedings of the Human Factors and Ergonomics Society Annual Meeting,50*, 1829–1832. https://doi.org/10.1177/154193120605001724

Chaparro, B. S., Shaikh, A. D., Chaparro, A., & Merkle, E. C. (2010). Comparing the legibility of six ClearType typefaces to Verdana and Times New Roman. *Information Design Journal,18*(1), 36–49. https://doi.org/10.1075/idj.18.1.04cha

Chernecky, C., Macklin, D., & Waller, J. (2006). Internet design preferences of patients with cancer. *Oncology Nursing Forum,33*(4), 787–792. https://doi.org/10.1188/06.ONF.787-792

Chicago manual of style (15th ed.). (2003). University of Chicago Press.

Chomsky, N. (1970). Remarks on nominalization. In R. A. Jacobs & P. S. Rosenbaum (Eds.), *Readings in English transformational grammar* (pp. 184–221). Ginn.

Chung, S. T. L. (2020). Reading in the presence of macular disease: A mini-review. *Ophthalmic and Physiological Optics,40*(2), 171–186. https://doi.org/10.1111/opo.12664

Click, J. W., & Stempel III, G. H. (1968). Reader response to newspaper front-page format. *Journal of Typographic Research,2*(2), 127–142.

Clough, J. (2015). A Roman legacy [Review of the book *The eternal letter*, edited by P. Shaw]. *Eye, 23*(90), 102–103. Retrieved June 30, 2021, from http://www.eyemagazine.com/review/article/a-roman-legacy.

Clough, J. (2020, September 16). *Letter-hunting in Italy/2: Historical inscriptions and more recent signs from all over the country.* Retrieved June 30, 2021, from https://articles.c-a-s-t.com/letter-hunting-in-italy-2-e7b51cd821a6.

Coghill, V. (1980). Can children read familiar words set in unfamiliar type? *Information Design Journal,1*(4), 254–260. https://doi.org/10.1075/idj.1.4.05cog

Cossu, G., Shankweiler, D., Liberman, I. Y., & Gugliotta, M. (1995). Visual and phonological determinants of misreadings in a transparent orthography. *Reading and Writing,7*(3), 237–256. https://doi.org/10.1007/BF02539523

Cowan, A. (1928). Test cards for the determination of the visual acuity: A review. *Archives of Ophthalmology,57*(May), 283–297.

Craig, J. (1971). *Designing with type: A basic course in typography* (S. E. Meyer, Ed.). Watson-Guptill.

Csilla, K. P., Péter, S., & Ádám, J. (2016). A papírról és képernyőről való olvasás és a talpas, illetve talpatlan betűtípusok hatása az elsajátításra [The impact of reading printed or monitor displayed texts and of the use of serif and sans-serif typeface on the knowledge acquisition process]. *Magyar Pszichológiai Szemle,71*(1), 91–108. https://doi.org/10.1556/0016.2016.71.1.5

Dale, N. (1902a). *The Dale readers: Infant reader* (new ed.). George Philip & Son.

Dale, N. (1902b). *Further notes on the teaching of English reading.* George Philip & Son.

Dale, N. (1903). *On the teaching of English reading* (2nd ed.). Dent.

Davidow, A. (2002). Computer screens are not like paper: Typography on the web. In R. Sassoon (Ed.), *Computers and typography 2*. Intellect.

Davidson, H. F. (1936). A study of the confusing letters *b*, *d*, *p*, and *q*. *The Pedagogical Seminary and Journal of Genetic Psychology,47*(2), 458–468. https://doi.org/10.1080/08856559.1935.10534056

Davis, R. C., & Smith, H. J. (1933). Determinants of feeling tone in type faces. *Journal of Applied Psychology,17*(6), 742–764. https://doi.org/10.1037/h0074491

De Bruijn, O., Spence, R., & Chong, M. Y. (2002). RSVP browser: Web browsing on small screen devices. *Personal and Ubiquitous Computing,6*(4), 245–252. https://doi.org/10.1007/s007790200024

De Lange, R. W., Esterhuizen, H. L., & Beatty, D. (1993). Performance differences between Times and Helvetica in a reading task. *Electronic Publishing, 6*(3), 241–248. Retrieved June 30, 2021, from http://cajun.cs.nott.ac.uk/compsci/epo/papers/volume6/issue3/rudi.pdf.

Deederer, C. (1968). Isometric ophthalmic chart. *American Journal of Ophthalmology,66*(4), 750. https://doi.org/10.1016/0002-9394(68)91304-4

Deederer, C. (1970). Sans serif letters for Snellen charts [Letter]. *Journal of the American Medical Association,211*(10), 1699.

Dillon, A., Kleinman, L., Bias, R., Choi, G. O., & Turnbull, D. (2004). Reading and searching digital documents: An experimental analysis of the effects of image quality on user performance and perceived effort. *Proceedings of the American Society for Information Science and Technology,41*(1), 267–273. https://doi.org/10.1002/meet.1450410131

Dillon, A., Kleinman, L., Choi, G. O., & Bias, R. (2006). Visual search and reading tasks using ClearType and regular displays: Two experiments. In *Proceedings of the SIGCHI Conference on Human Factors in Computing Systems* (pp. 503–511). Association for Computing Machinery. https://doi.org/10.1145/1124772.1124849

Dreyfus, J. (1985). A turning point in type design. *Visible Language, 19*(1), 11–22. Retrieved June 30, 2021, from https://s3-us-west-2.amazonaws.com/visiblelanguage/pdf/19.1/a-turning-point-in-type-design.pdf.

Dyson, M. C. (2005). How do we read text on screen? In H. van Oostendorp, L. Breure, & A. Dillon (Eds.) *Creation, use, and deployment of digital information* (pp. 279–306). Routledge. https://doi.org/10.4324/9781410613035

Earnest, W. J. (2003). *Developing strategies to evaluate the effective use of electronic presentation software in communication education* [Doctoral dissertation, University of Texas at Austin]. https://repositories.lib.utexas.edu/bitstream/handle/2152/557/earnestwj039.pdf

English, E. (1944). A study of the readability of four newspaper headline types. *Journalism Quarterly,21*(3), 217–229. https://doi.org/10.1177/107769904402100303

Errando Oyonarte, C. L. (2011a). Nuevo aspect para la *Revista Española de Anestesiología y Reanimación* [New look for *Revista Española de Anestesiología y Reanimación*] [Editorial]. *Revista Española De Anestesiología y Reanimación,58*(2), 69. https://doi.org/10.1016/s0034-9356(11)70001-9

Errando Oyonarte, C. L. (2011b). Respuesta a la carta al director "Palo seco" [Response to the letter to the editor "Dry stick"] [Letter]. *Revista Española De Anestesiología y Reanimación,58*(5), 326. https://doi.org/10.1016/S0034-9356(11)70073-1

Estey, A., Jeremy, P., & Jones, M. (1990). Developing printed materials for patients with visual deficiencies. *Journal of Ophthalmic Nursing and Technology,9*(6), 247–249.

Fox, D., Chaparro, B. S., & Merkle, E. (2007a). Examining legibility of the letter "e" and the number "0" using classification tree analysis. *Usability News*, Vol. 9, No. 2. https://web.archive.org/web/20080712041546/http://www.surl.org/usabilitynews/92/legibility.asp

Fox, D., Shaikh, A. D., & Chaparro, B. S. (2007). The effect of typeface on the perception of onscreen documents. *Proceedings of the Human Factors and Ergonomics Society Annual Meeting,51*(5), 464–468. https://doi.org/10.1177/154193120705100508

Fox, D., Chaparro, B. S., & Merkle, E. (2008). Examining the onscreen legibility of the number "0" and the number "1." *Proceedings of the Human Factors and Ergonomics Society Annual Meeting,52*(6), 547–551. https://doi.org/10.1177/154193120805200608

Fox, J. G. (1963). A comparison of Gothic Elite and Standard Gothic typefaces. *Ergonomics,6*(2), 193–198. https://doi.org/10.1080/00140136308930689

Gallagher, T. J., & Jacobson, W. S. (1993). The typography of environmental impact statements: Criteria, evaluation, and public participation. *Environmental Management,17*(1), 99–109. https://doi.org/10.1007/BF02393798

Garcia, M. L., & Caldera, C. I. (1996). The effect of color and typeface on the readability of on-line text. *Computers and Industrial Engineering,31*(1–2), 519–524. https://doi.org/10.1016/0360-8352(96)00189-1

Garon, J. E. (1999). Presentation skills for the reluctant speaker. *Clinical Laboratory Management Review,13*(6), 372–385.

Gasser, M., Boeke, J., Haffernan, M., & Tan, R. (2005). The influence of font type on information recall. *North American Journal of Psychology, 7*(2), 181–188. Retrieved June 30, 2021, from https://www.researchgate.net/publication/237229931_The_Influence_of_Font_Type_on_Information_Recall.

Gilbert, L. C. (1959a). Saccadic movements as a factor in visual perception in reading. *Journal of Educational Psychology,50*(1), 15–19. https://doi.org/10.1037/h0040752

Gilbert, L. C. (1959b). Speed of processing verbal stimuli and its relation to reading. *Journal of Educational Psychology,50*(1), 8–14. https://doi.org/10.1037/h0045592

Goikoetxea, E. (2006). Reading errors in first- and second-grade readers of a shallow orthography: Evidence from Spanish. *British Journal of Educational Psychology,76*(2), 333–350. https://doi.org/10.1348/000709905X52490

González-Rodriguez, R. (2011). Palo seco [Dry stick] [Letter]. *Revista Española De Anestesiología y Reanimación,58*(5), 325–326. https://doi.org/10.1016/S0034-9356(11)70072-X

Gosse, R. (1999). *Adapting Hvistendahl's and Kahl's typographic legibility study to the World Wide Web* (Unpublished master's thesis). Ball State University.

Gould, J. D., Alfaro, L., Finn, R., Haupt, B., & Minuto, A. (1987). Reading from CRT displays can be as fast as reading from paper. *Human Factors,29*(5), 497–517. https://doi.org/10.1177/001872088702900501

Grant, M. M., & Branch, R. M. (2000, October). *Performance differences between serif fonts and sans serif fonts in an on-screen reading task* [Paper presentation]. International Visual Literacy Association annual conference, Ames, IA.

Gray, N. (1960). Sans serif and other experimental inscribed lettering of the early Renaissance. *Motif,5*(Autumn), 67–76.

Griffing, H., & Franz, S. I. (1896). On the conditions of fatigue in reading. *Psychological Review,3*(5), 513–530. https://doi.org/10.1037/h0075858

Griffiths, G. W. (2020). The Comparative Rate of Reading Test (CRST): The importance of comparing the rate of individual character recognition (static tracking speed) in serif and non-serif fonts. *EC Ophthalmology, 11*(8), 6–17. Retrieved June 30, 2021, from https://www.ecronicon.com/ecop/pdf/ECOP-11-00677.pdf.

Grooters, L. E. (1972). *The relationship of letter style, letter size, and viewing distance to the readability of transparent visuals* [Doctoral dissertation, University of Oklahoma]. Retrieved June 30, 2021, from https://shareok.org/handle/11244/3428.

Gugerty, L., Tyrrell, R. A., Aten, T. R., & Edmonds, K. A. (2004). The effects of subpixel addressing on users' performance and preferences during reading-related tasks. *ACM Transactions on Applied Perceptions,1*(2), 81–101. https://doi.org/10.1145/1024083.1024084

Hamai, M. (1986). *An introduction to typography* (General Graphic Arts, Lesson Plan No. 5) (ED283547). ERIC. https://files.eric.ed.gov/fulltext/ED283547.pdf

Hargreaves, A. (2008). When sight can be tiring and painful. BBC News. http://news.bbc.co.uk/1/hi/health/7661998.stm

Harris, J. (1973). Confusions in letter recognition. *Printing Technology,17*(2), 29–34.

Hartley, J., Fraser, S., & Burnhill, P. (1975). Some observations on the reliability of measures used in reading and typographic research. *Journal of Reading Behavior,7*(3), 283–286. https://doi.org/10.1080/10862967509547146

Hartley, J., & Rooum, D. (1983). Sir Cyril Burt and typography: A re-evaluation. *British Journal of Psychology,74*(2), 203–212. https://doi.org/10.1111/j.2044-8295.1983.tb01856.x

Haskins, J. B. (1958). Testing suitability of typefaces for editorial subject-matter. *Journalism Quarterly,35*(2), 186–194. https://doi.org/10.1177/107769905803500205

Haskins, J. B., & Flynne, L. P. (1974). Effect of headline typeface variation on reading interest. *Journalism Quarterly,51*(4), 677–682. https://doi.org/10.1177/107769907405100415

Haugen, T. T. (2010). *Seeking visual clarity: An examination of font legibility and visual presentation for elementary-level special education students* (UMI No. AAI3396931) [Doctoral dissertation, University of Minnesota]. ProQuest Dissertations and Theses Global.

Haw, C. (2017). *Accessible information: Collaborating with people with aphasia to develop an evidence-based template for health information* [Unpublished doctoral dissertation, University of Sheffield]. https://etheses.whiterose.ac.uk/17991/1/CarolineHaw_Thesis_August2017.pdf

Hawking, S. W., & Penrose, R. (1999). *A tér és az idő természete* [The nature of space and time] (E. Both, Trans.). Akkord. (Original work published 1996)

Hearnshaw, L. S. (1979). *Cyril Burt, psychologist*. Hodder & Stoughton.

Hedges, L. V., & Olkin, I. (1985). *Statistical methods for meta-analysis*. Academic Press.

Hedlich, C., Barstow, E., & Vogtle, L. K. (2018). Age-related macular degeneration and reading performance: Does font style make a difference? *Journal of Visual Impairment and Blindness,112*(4), 398–402. https://doi.org/10.1177/0145482X1811200406

Herbert, R., Gregory, E., & Haw, C. (2019). Collaborative design of accessible information with people with aphasia. *Aphasiology,33*(12), 1504–1530. https://doi.org/10.1080/02687038.2018.1546822

Hering, E. (1879). Über Muskelgeräusche des Auges [Concerning muscle noise of the eye]. *Sitzberichte der kaiserlichen Akademie der Wissenschaften in Wien, Mathematisch-naturwissenschaftliche Klasse, 79*(3), 137–154. Retrieved June 30, 2021, from https://wellcomecollection.org/works/chjc6zwb.

Hetherington, R. (1954). The Snellen Chart as a test of visual acuity. *Psychologische Forschung,24*(4), 349–357. https://doi.org/10.1007/BF00422033

Hofstätter, P. R. (1966). Objektive Methoden zur Erfassung von Anmutungsqualitäten [Objective methods for understanding perceptual qualities]. *Exakte Ästhetik,3*(4), 47–65.

Hojjati, N., & Muniandy, B. (2014). The effects of font type and spacing of text for online readability and performance. *Contemporary Educational Technology, 5*(2), 161–174. https://doi.org/10.30935/cedtech/6122

Holleran, P. A. (1992). An assessment of font preferences for screen-based text display. In A. Monk, D. Diaper, & M. D. Harrison (Eds.), *People and computers VII: Proceedings of the HCI'92 Conference* (British Computer Society Conference Series 5, pp. 449–461). Cambridge University Press.

Hutner, N., Duboff, J. M., Oscar-Berman, M., & Mueller, S. (1999). Comparing visual perception on conventional cabinet tachistoscopes and computer monitor tachistoscopes. *Behavior Research Methods, Instruments, and Computers,31*(3), 400–409. https://doi.org/10.3758/BF03200718

Hvistendahl, J. K., & Kahl, M. R. (1975). Roman v. sans serif body type: Readability and reader preference. *News Research Bulletin*, No. 2, pp. 1–12 (ED145419). ERIC. https://files.eric.ed.gov/fulltext/ED145419.pdf

Hwang, W. S., Lee, D. C., & Lee, S. D. (1997). An experimental study on search speed and error rate according to Korean letter size and font on search task with VDT. *Journal of the Ergonomics Society of Korea, 16*(2), 29–38. http://jesk.or.kr/archive/detail/320

Ing, E., Celo, E., Ing, R., Weisbrod, L., & Ing, M. (2017). Quantitative analysis of the text and graphic content in ophthalmic slide presentations. *Canadian Journal of Ophthalmology,52*(2), 171–174. https://doi.org/10.1016/j.jcjo.2016.11.029

Jainta, S., Jaschinski, W., & Wilkins, A. J. (2010). Periodic letter strokes within a word affect fixation disparity during reading. *Journal of Vision, 10*(13), Article No. 2. https://doi.org/10.1167/10.13.2

Jelfs, A., & Richardson, J. T. E. (2013). The use of digital technologies across the adult life span in distance education. *British Journal of Educational Technology,44*(2), 338–351. https://doi.org/10.1111/j.1467-8535.2012.01308.x

Jorgensen, B. (2003). Baby Boomers, Generation X and Generation Y? Policy implications for defence forces in the modern era. *Foresight,5*(4), 41–49. https://doi.org/10.1108/14636680310494753

Josephson, S. (2008). Keeping your readers' eyes on the screen: An eye-tracking study comparing sans serif and serif typefaces. *Visual Communication Quarterly,15*(1–2), 67–79. https://doi.org/10.1080/15551390801914595

Juni, S., & Gross, J. S. (2008). Emotional and persuasive perception of fonts. *Perceptual and Motor Skills,106*(1), 35–42. https://doi.org/10.2466/pms.106.1.35-42

Kaspar, K., Wehlitz, T., von Knobelsdorff, S., Wulf, T., & von Saldern, M. A. O. (2015). A matter of font type: The effect of serifs on the evaluation of scientific abstracts. *International Journal of Psychology,50*(5), 372–378. https://doi.org/10.1002/ijop.12160

Kastl, A. J., & Child, I. L. (1968). Emotional meaning of four typographical variables. *Journal of Applied Psychology, 52*(6, Pt. 1), 440–446. https://doi.org/10.1037/h0026506

Kempson, E., & Moore, N. (1994). *Designing public documents: A review of research*. Policy Studies Institute.

Kennedy, H. (1954). Reversals, reversals, reversals. *Journal of Experimental Education,23*(2), 161–170. https://doi.org/10.1080/00220973.1954.11010502

Kerr, J. (1926). *Fundamentals of school health*. Allen & Unwin. Retrieved June 30, 2021, from http://hdl.handle.net/2027/uc1.$b67827.

Kim, M., Park, S.-H., Ahn, S. H., Choi, M. B., & Yun, M.-H. (2015). Analysis of font legibility for Korean serif and sans serif types in printed and displayed settings. *Proceedings of the Human Factors and Ergonomics Society,59*(1), 1774–1777. https://doi.org/10.1177/1541931215591383

Kinross, R. (1992). *Modern typography: An essay in critical history*. Hyphen Press.

Koch, B. E. (2012). Emotion in typographic design: An empirical examination. *Visible Language, 46*(3), 206–227. Retrieved June 30, 2021, from https://s3-us-west-2.amazonaws.com/visiblelanguage/pdf/V46N3_2012_E.pdf.

Kong, Y.-K., Lee, I., Jung, M.-C., & Song, Y.-W. (2011). The effects of age, viewing distance, display type, font type, colour contrast and number of syllables on the legibility of Korean characters. *Ergonomics,54*(5), 453–465. https://doi.org/10.1080/00140139.2011.568635

Krulee, G. K., & Novy, F. (1986). Word processing and effects of variability in type fonts. *Perceptual and Motor Skills,62*(3), 999–1010. https://doi.org/10.2466/pms.1986.62.3.999

Kullmann, D. M. (2015). Editorial. *Brain,138*(1), 1. https://doi.org/10.1093/brain/awu366

Kuster, S. M., van Weerdenburg, M., Gompel, M., & Bosman, A. M. T. (2018). Dyslexie font does not benefit reading in children with or without dyslexia. *Annals of Dyslexia,68*(1), 25–42. https://doi.org/10.1007/s11881-017-0154-6

Lamare, M. (1892). Des mouvements des yeux dans la lecture [Eye movements in reading]. *Bulletins et Mémoires de la Société Française d'Ophtalmologie, 10*, 354–364. Retrieved June 30, 2021, from https://pure.mpg.de/rest/items/item_2352590/component/file_2628648/content.

Legge, G. E., & Bigelow, C. A. (2011). Does print size matter for reading? A review of findings from vision science. *Journal of Vision, 11*(5), Article No. 8. https://doi.org/10.1167/11.5.8

Legros, L. A. (1922). *A note on the legibility of printed matter*. His Majesty's Stationery Office.

Lenze, J. S. (1991). Serif vs. sans serif type fonts: A comparison based on reader comprehension. In D. G. Beauchamp, J. C. Baca, & R. A. Braden (Eds.), *Investigating Visual Literacy: Selected Readings from the 22nd Annual Conference of the International Visual Literacy Association* (pp. 93–98). International Visual Literacy Association (ED352051). ERIC. https://files.eric.ed.gov/fulltext/ED352051.pdf

References

Lightfoot, C. (2009). Roman inscriptions. In *Heilbrunn Timeline of Art History*. The Metropolitan Museum of Art. Retrieved June 30, 2021, from http://www.metmuseum.org/toah/hd/insc/hd_insc.htm.

Ling, J., & van Schaik, P. (2006). The influence of font type and line length on visual search and information retrieval in web pages. *International Journal of Human-Computer Studies, 64*(5), 395–404. https://doi.org/10.1016/j.ijhcs.2005.08.015

Liversedge, S. P., White, S. J., Findlay, J. M., & Rayner, K. (2006). Binocular coordination of eye movements during reading. *Vision Research, 46*(15), 2363–2374. https://doi.org/10.1016/j.visres.2006.01.013

Lockhead, G. R., & Crist, W. B. (1980). Making letters distinctive. *Journal of Educational Psychology, 72*(4), 483–493. https://doi.org/10.1037/0022-0663.72.4.483

Luckiesh, M., & Moss, F. K. (1935). Visibility: Its measurement and significance in seeing. *Journal of the Franklin Institute, 220*(4), 431–466. https://doi.org/10.1016/S0016-0032(35)90130-2

Luckiesh, M., & Moss, F. K. (1937). The visibility of various type faces. *Journal of the Franklin Institute, 223*(1), 77–82. https://doi.org/10.1016/S0016-0032(37)90585-4

Luckiesh, M., & Moss, F. K. (1942). *Reading as a visual task*. Van Nostrand.

Lund, O. (1999). *Knowledge construction in typography: The case of legibility research and the legibility of sans serif typefaces* [Doctoral dissertation, University of Reading]. https://ethos.bl.uk/OrderDetails.do?uin=uk.bl.ethos.301973

Maccoby, E. E., & Jacklin, C. N. (1974). *The psychology of sex differences*. Stanford University Press.

Mach, E. (1897). *Contributions to the analysis of the sensations* (C. M. Williams, Trans.). Open Court. (Original work published 1886). https://archive.org/details/contributionsto00machgoog/page/n7/mode/2up

Mackiewicz, J. (2007). Audience perceptions of fonts in projected PowerPoint slides. *Technical Communication, 54*(3), 295–307. Retrieved June 30, 2021, from https://ieeexplore.ieee.org/document/4114184.

Mackiewicz, J., & Moeller, R. (2004). Why people perceive typefaces to have different personalities. *Proceedings of the International Professional Communication Conference, 2004*, 304–313. https://doi.org/10.1109/IPCC.2004.1375315

Marinus, E., Mostard, M., Segers, E., Schubert, T. M., Madelaine, A., & Wheldall, K. (2016). A special font for people with dyslexia: Does it work and if so, why? *Dyslexia, 22*(3), 233–244. https://doi.org/10.1002/dys.1527

Mátrai, R., & Kosztyán, Z. T. (2014). How can we improve text comprehension on web pages? *Global Journal on Technology, 5*(1), 66–72. http://archives.un-pub.eu/index.php/P-ITCS/article/view/3047/2457

McCarthy, M. S., & Mothersbaugh, D. L. (2002). Effects of typographic factors in advertising-based persuasion: A general model and initial empirical tests. *Psychology and Marketing, 19*(7–8), 663–692. https://doi.org/10.1002/mar.10030

McLean, R. (1980). *The Thames and Hudson manual of typography*. Thames & Hudson.

McVey, G. F. (1985). Legibility in film-based and television display systems. *Technical Communication, 32*(4), 21–28. https://www.jstor.org/stable/i40120370

Merriam-Webster's manual for writers and editors. (1998). Merriam-Webster. https://archive.org/details/merriamwebstersm0000merr

Mertens, S., & Baethge, C. (2011). The virtues of correct citation: Careful referencing is important but is often neglected even in peer reviewed articles. *Deutsches Ärzteblatt International, 108*(33), 550–552. https://doi.org/10.3238/arztebl.2011.0550

Misanchuk, E. R. (1989). Learner/user preferences for fonts in microcomputer screen displays. *Canadian Journal of Educational Communication, 18*(3), 193–205. https://doi.org/10.21432/T2731K

Mizrachi, D. (2015). Undergraduates' academic reading format preferences and behaviors. *Journal of Academic Librarianship, 41*(3), 301–311. https://doi.org/10.1016/j.acalib.2015.03.009

Mollen, J. D., & Polden, P. G. (1978). On the time constants of tachistoscopes. *Quarterly Journal of Experimental Psychology,30*(3), 555–568. https://doi.org/10.1080/00335557843000133

Moret-Tatay, C., & Perea, M. (2011). Do serifs provide an advantage in the recognition of written words? *Journal of Cognitive Psychology,23*(5), 619–624. https://doi.org/10.1080/20445911.2011.546781

Moriarty, S. E., & Scheiner, E. C. (1984). A study of close-set text type. *Journal of Applied Psychology,69*(4), 700–702. https://doi.org/10.1037/0021-9010.69.4.700

Morison, S. (1959). Introduction. In C. Burt, *A psychological study of typography* (pp. ix–xix). Cambridge University Press.

Morris, R. A., Aquilante, K., Yager, D., & Bigelow, C. (2002). Serifs slow RSVP reading at very small sizes, but don't matter at larger sizes. *Society for Information Display International Symposium Digest of Technical Papers,33*(1), 244–247. https://doi.org/10.1889/1.1830242

Morrison, G. R. (1986). Communicability of the emotional connotation of type. *Educational Communication and Technology, 34*(4), 235–244. https://www.jstor.org/stable/30218198

Mosley, J. (1999). *The nymph and the grot*. Friends of the St. Bride Printing Library.

Mosley, J. (2007, January 6). The nymph and the grot, an update [Web log post]. Retrieved June 30, 2021, from http://typefoundry.blogspot.com/2007/01/nymph-and-grot-update.html.

Muter, P., & Mauretto, P. (1991). Reading and skimming from computer screens and books. *Behaviour and Information Technology,10*(1), 257–266. https://doi.org/10.1080/01449299108924288

Myung, R. (2003). Conjoint analysis as a new methodology for Korean typography guideline in Web environments. *International Journal of Industrial Ergonomics,32*(5), 341–348. https://doi.org/10.1016/S0169-8141(03)00074-X

Nersveen, J., Kvitle, A. K., & Johansen, E. A. (2018). Legibility in print text for people with impaired vision. In G. Craddock, C. Doran, L. McNutt, & D. Rice (Eds.), *Studies in health technology and informatics: Vol. 256. Transforming our world through design, diversity and education* (pp. 862–869). IOS Press. https://doi.org/10.3233/978-1-61499-923-2-862

Nolan, C. Y. (1959). Readability of large types: A study of type size and type styles. *International Journal for the Education of the Blind,9*(1), 41–44.

Noonan, E., & Bjørndal, A. (2010). The Campbell collaboration. *Cochrane Database of Systematic Reviews,9*, 1–2. https://doi.org/10.1002/14651585.ED000011

Nutt, D. J. (2004). Some important changes [Editorial]. *Journal of Psychopharmacology,18*(1), 5. https://doi.org/10.1177/0269881104043034

Osgood, C. E., May, W. H., & Miron, M. S. (1975). *Cross-cultural universals of affective meaning*. University of Illinois Press.

Osgood, C. E., Suci, G. J., & Tannenbaum, P. H. (1957). *The measurement of meaning*. University of Illinois Press.

Ovink, G. W. (1938). *Legibility, atmosphere-value and forms of printing types*. Sijthoff. Retrieved June 30, 2021, from https://resolver.kb.nl/resolve?urn=MMKB06:000008331:pdf.

Oxford University Press. (n.d.). Font. In *Oxford English Dictionary*. Retrieved June 30, 2021, from https://www.oed.com/view/Entry/72622.

Park, K.-S., Ann, S. H., Kim, C.-H., Park, M., & Lee, S.-S. (2008). The effects of Hangul font and character size on the readability of PDA. In O. Gervasi, B. Murgante, A. Laganà, D. Taniar, Y. Mun, & M. L. Gavrilova (Eds.), *Computational Science and its Applications: ICCSA 2008* (Lecture Notes in Computer Science, No. 5073, pp. 601–614). Springer. https://doi.org/10.1007/978-3-540-69839-5_45

Paterson, D. G., & Tinker, M. A. (1932). Studies of typographical factors influencing speed of reading: X. Style of type face. *Journal of Applied Psychology, 16*(6), 605–613. https://doi.org/10.1037/h0070644

Pedró, F. (2009). *New millennium learners in higher education: Evidence and policy implications*. Organisation for Economic Co-operation and Development, Centre for Educational Research and Innovation. Retrieved June 30, 2021, from https://www.researchgate.net/publication/242554689_New_Millennium_Learners_in_Higher_Education_Evidence_and_Policy_Implications.

Perea, M. (2013). Why does the APA recommend the use of serif fonts? *Psicothema,25*(1), 13–17. https://doi.org/10.7334/psicothema2012.141

Perea, M., Panadero, V., Moret-Tatay, C., & Gómez, P. (2012). The effects of inter-letter spacing in visual-word recognition: Evidence with young normal readers and developmental dyslexics. *Learning and Instruction,22*(6), 420–430. https://doi.org/10.1016/j.learninstruc.2012.04.001

Phillips, R. M. (1976). *The interacting effects of letter style, letter stroke-width and letter size on the legibility of projected high contrast lettering* (UMI No. 7703361) [Doctoral dissertation, Indiana University]. ProQuest Dissertations and Theses Global.

Pittman, B. (1976). Style of type and the reading comprehension of learning disabled and normal pupils. *Journal of Research and Development in Education,9*, 114–115.

Plass, B., & Yager, D. (1995). Presentation of video-based text to low vision readers: Rapid serial visual presentation vs. full-page reading [Abstract]. *Investigative Ophthalmology and Visual Science, 36*(4), S671. https://iovs.arvojournals.org/issues.aspx?issueid=933581&journalid=177#issueid=933581

Poffenberger, A. T., & Franken, R. B. (1923). A study of the appropriateness of type faces. *Journal of Applied Psychology,7*(4), 312–319. https://doi.org/10.1037/h0071591

Poncelet, G. M., & Proctor, L. F. (1993). Design and development factors in the production of hypermedia-based courseware. *Canadian Journal of Educational Communication, 22*(2), 91–111. https://doi.org/10.21432/T2B898

Poole, A. (2012, March 8). Fighting bad typography research [Web log post]. Retrieved June 30, 2021, from http://alexpoole.info/blog/fighting-bad-typography-research/

Popp, H. M. (1964). Visual discrimination of alphabet letters. *The Reading Teacher, 17*(4), 221–226. https://www.jstor.org/stable/20197756

Popper, K. R. (1959). *The logic of scientific discovery*. Hutchinson.

Popper, K. R. (1962). *Conjectures and refutations: The growth of scientific knowledge*. Routledge & Kegan Paul.

Potter, M. C. (1949). *Perception of symbol orientation and early reading success* (Contributions to Education, No. 939). New York: Columbia University, Teachers College.

Poulton, E. C. (1965). Letter differentiation and rate of comprehension in reading. *Journal of Applied Psychology,49*(5), 358–362. https://doi.org/10.1037/h0022461

Poulton, E. C. (1972). Size, style, and vertical spacing in the legibility of small typefaces. *Journal of Applied Psychology,56*(2), 156–161. https://doi.org/10.1037/h0032670

Poulton, E. C., & Brown, C. H. (1967). Memory after reading aloud and reading silently. *British Journal of Psychology,58*(3–4), 219–222. https://doi.org/10.1111/j.2044-8295.1967.tb01076.x

Powell, S. L., & Trice, A. D. (2020). The impact of a specialized font on the reading performance of elementary children with reading disability. *Contemporary School Psychology,24*(1), 34–40. https://doi.org/10.1007/s40688-019-00225-4

Prensky, M. (2001). Digital natives, digital immigrants Part 1. *On the Horizon,9*(5), 1–6. https://doi.org/10.1108/10748120110424816

Prince, J. H. (1967). Printing for the visually handicapped. *Journal of Typographic Research,1*(1), 31–47.

Pyke, R. L. (1926). *The legibility of print* (Medical Research Council, Special Report Series No. 110). His Majesty's Stationery Office.

Raban, B. (1984). Survey of teachers' opinions: Children's books and handwriting styles. In D. Dennis (Ed.), *Reading: Meeting children's special needs* (pp. 123–129). Heinemann.

Raymaker, D. M., Kapp, S. K., McDonald, K. E., Weiner, M., Ashkenazy, E., & Nicolaidis, C. (2019). Development of the AASPIRE web accessibility guidelines for autistic web users. *Autism in Adulthood,1*(2), 146–157. https://doi.org/10.1089/aut.2018.0020

Redich, J. (2012). *Letting go of words: Writing web content that works* (2nd ed.). Morgan Kaufmann.

Rello, L., & Baeza-Yates, R. (2013). Good fonts for dyslexia. In *ASSETS'13: Proceedings of the 15th International ACM SIGACCESS Conference on Computers and Accessibility*, Article No. 14. https://doi.org/10.1145/2513383.2513447

Reynolds, L. (1979). Legibility studies: Their relevance to present-day documentation methods. *Journal of Documentation,35*(4), 307–340. https://doi.org/10.1108/eb026686

Richards, O. W. (1965). A comparison of acuity test letters with and without serifs. *American Journal of Optometry and Archives of the American Academy of Optometry,42*(10), 589–592. https://doi.org/10.1097/00006324-196510000-00004

Richards, O. W. (1978). A comparison of acuity test letters with and without serifs: Final report. *American Journal of Optometry and Archives of the American Academy of Optometry,55*(6), 407–408.

Ripoli, J. C. (2015). Font legibility in first year primary students. *Infancia y Aprendizaje/journal for the Study of Education and Development,38*(3), 600–616. https://doi.org/10.1080/02103702.2015.1054668

Robinson, D. O., Abbamonte, M., & Evans, S. H. (1971). Why serifs are important: The perception of small print. *Visible Language,4*(4), 353–359.

Roethlein, B. E. (1912). The relative legibility of different faces of printing types. *American Journal of Psychology,* 23(1), 1–36. https://doi.org/10.2307/1413112. (Also published as Roethlein, B. E. (1912). The relative legibility of different faces of printing types. *Publications of the Clark University Library, 3*(1), 1–41. Retrieved June 30, 2021, from https://babel.hathitrust.org/cgi/pt?id=hvd.32044080330855&view=1up&seq=11).

Rose, T. A., Worrall, L. E., Hickson, L. M., & Hoffmann, T. C. (2011). Aphasia friendly written health information: Content and design characteristics. *International Journal of Speech-Language Pathology,13*(4), 335–347. https://doi.org/10.3109/17549507.2011.560396

Rowe, C. L. (1982). The connotative dimensions of selected display typefaces. *Information Design Journal,3*(1), 30–37. https://doi.org/10.1075/idj.3.1.03row

Rubin, G. S., Feely, M., Perera, S., Ekstrom, K., & Williamson, E. (2006). The effect of font and line width on reading speed in people with mild to moderate vision loss. *Ophthalmic and Physiological Optics,26*(6), 545–554. https://doi.org/10.1111/j.1475-1313.2006.00409.x

Rubin, G. S., & Turano, K. (1992). Reading without saccadic eye movements. *Vision Research,32*(5), 895–902. https://doi.org/10.1016/0042-6989(92)90032-E

Russell-Minda, E., Jutai, J., Strong, G., Campbell, K., Gold, D., Pretty, L., & Wilmot, L. (2006). *An evidence-based review of the research on typeface legibility for readers with low vision.* Canadian National Institute for the Blind. Retrieved June 30, 2021, from https://studylib.net/doc/7324361/clear-print-full-review.

Russell-Minda, E., Jutai, J. W., Strong, J. G., Campbell, K. A., Gold, D., Pretty, L., & Wilmot, L. (2007). The legibility of typefaces for readers with low vision: A research review. *Journal of Visual Impairment and Blindness,101*(7), 402–415. https://doi.org/10.1177/0145482X0710100703

Sanocki, T. (1987). Visual knowledge underlying letter perception: Font-specific, schematic tuning. *Journal of Experimental Psychology: Human Perception and Performance,13*(2), 267–278. https://doi.org/10.1037/0096-1523.13.2.267

Sanocki, T. (1988). Font regularity constraints on the process of letter recognition. *Journal of Experimental Psychology: Human Perception and Performance,14*(3), 472–480. https://doi.org/10.1037/0096-1523.14.3.472

Sassoon, R. (1993). Through the eyes of a child: Perception and type design. In R. Sassoon (Ed.), *Computers and typography* (pp. 150–177). Intellect.

Savory, P., Crowe, J. V., & Hallbeck, M. S. (2012). Focus group analysis of hand-held radiation detector design. *International Journal of Industrial Ergonomics,42*(1), 17–24. https://doi.org/10.1016/j.ergon.2011.11.003

Schiffman, H. R. (2000). *Sensation and perception: An integrated approach* (5th ed.). Wiley.

Schiller, G. (1935). An experimental study of the appropriateness of color and type in advertising. *Journal of Applied Psychology,19*(6), 652–664. https://doi.org/10.1037/h0056090

Schriver, K. A. (1997). *Dynamics in document design: Creating text for readers.* Wiley.

Schweinberger, S. R., Ramsay, A. L., & Kaufmann, J. M. (2006). Hemispheric asymmetries in font-specific and abstractive priming of written personal names: Evidence from event-related potentials. *Brain Research,1117*(1), 195–205. https://doi.org/10.1016/j.brainres.2006.08.070

Seale, J. (2006). *E-learning and disability in higher education: Accessibility research and practice.* Routledge.

Shaikh, A. D. (2004). Paper or pixels: What are people reading online? *Usability News*, Vol. 6, No. 2. Retrieved June 30, 2021, from https://web.archive.org/web/20080807173440/http://www.surl.org/usabilitynews/62/online_reading.asp.

Shaikh, A. D. (2007). *Psychology of onscreen type: Investigations regarding typeface personality, appropriateness, and impact on document perception* (Doctoral dissertation). Wichita State University, Wichita, KS. Retrieved June 30, 2021, from https://soar.wichita.edu/handle/10057/1109.

Shaikh, A. D., & Chaparro, B. S. (2004). A survey of online reading habits of Internet users. *Proceedings of the Human Factors and Ergonomics Society Annual Meeting,48*(5), 875–879. https://doi.org/10.1177/154193120404800528

Shaikh, A. D., Chaparro, B. S., & Fox, D. (2006). Perception of fonts: Perceived personality traits and uses. *Usability News*, Vol. 8, No. 1. Retrieved June 30, 2021, from https://web.archive.org/web/20080725064702/http://www.surl.org/usabilitynews/81/PersonalityofFonts.asp.

Shaw, A. (1969). *Print for partial sight: A report to the Library Association Sub-committee on Books for Readers with Defective Sight.* The Library Association.

Shaw, S. C. K., & Anderson, J. L. (2017). Twelve tips for teaching medical students with dyslexia. *Medical Teaching,39*(7), 686–690. https://doi.org/10.1080/0142159X.2017.1302080

Sheedy, J. E., & McCarthy, M. (1994). Reading performance and visual comfort with scale to gray compared with black-and-white scanned print. *Displays,15*(1), 27–30. https://doi.org/10.1016/0141-9382(94)90040-X

Sheedy, J., Tai, Y.-C., Subbaram, M., Gowrisankaran, S., & Hayes, J. (2008). ClearType sub-pixel text rendering: Preference, legibility and reading performance. *Displays,29*(2), 138–151. https://doi.org/10.1016/j.displa.2007.09.016

Singer, L. M., & Alexander, P. A. (2017). Reading across mediums: Effects of reading digital and print texts on comprehension and calibration. *Journal of Experimental Education,85*(1), 155–172. https://doi.org/10.1080/00220973.2016.1143794

Skilton, A., Bowell, W., Prince, K., Francome-Wood, P., & Moosajee, M. (2018). Overcoming barriers to the involvement of deafblind people in conversations about research: Recommendations from individuals with Usher syndrome. *Research Involvement and Engagement*, 4, Article No. 40. https://doi.org/10.1186/s40900-018-0124-0

Slattery, T. J., & Rayner, K. (2010). The influence of text legibility on eye movements during reading. *Applied Cognitive Psychology,24*(8), 1129–1148. https://doi.org/10.1002/acp.1623

Sloan, L. L. (1959). New test charts for the measurement of visual acuity at far and near distances. *American Journal of Ophthalmology,48*(6), 807–813. https://doi.org/10.1016/0002-9394(59)90626-9

Smither, J.A.-A., & Braun, C. C. (1994). Readability of prescription drug labels by older and younger adults. *Journal of Clinical Psychology in Medical Settings,1*(2), 149–159. https://doi.org/10.1007/BF01999743

Snellen, H. (1862). *Letterproeven, tot bepaling der gezigtsscherpte* [Test types for the determination of the acuteness of vision]. Van de Weijer.

Soleimani, H., & Mohammadi, E. (2012). The effect of text typographical features on legibility, comprehension, and retrieval of EFL learners. *English Language Teaching,5*(8), 207–216. https://doi.org/10.5539/elt.v5n8p207

Song, H., & Schwarz, N. (2008a). Fluency and the detection of misleading questions: Low processing fluency attenuates the Moses illusion. *Social Cognition,26*(6), 791–799. https://doi.org/10.1521/soco.2008.26.6.791

Song, H., & Schwarz, N. (2008b). If it's hard to read, it's hard to do: Processing fluency affects effort prediction and motivation. *Psychological Science,19*(10), 986–988. https://doi.org/10.1111/j.1467-9280.2008.02189.x

Stiff, P. (2005). Sans serif and other experimental inscribed lettering of the early Renaissance. In P. Stiff (Ed.), *Typography papers 6* (pp. 66–114). Hyphen Press.

Stone, D. B., Fisher, S. K., & Eliot, J. (1999). Adults' prior exposure to print as a predictor of the legibility of text on paper and laptop computer. *Reading and Writing,11*(1), 1–28. https://doi.org/10.1023/A:1007993603998

Strauss, W., & Howe, N. (1991). *Generations: The history of America's future, 1584 to 2069*. Quill.

Suen, C. Y., & Komoda, M. K. (1986). Legibility of digital type-fonts and comprehension in reading. In J. C. van Vliet (Ed.), *Text processing and document manipulation* (pp. 178–187). Cambridge University Press.

Svensson, E. (2019). *Seriffers påverkan på läshastigheten* [The impact of serifs on reading speed] [Unpublished master's thesis]. Linnaeus University. Retrieved June 30, 2021, from http://lnu.diva-portal.org/smash/get/diva2:1326164/FULLTEXT01.pdf.

Tannenbaum, P. H., Jacobson, H. K., & Norris, E. L. (1964). An experimental investigation of typeface connotations. *Journalism Quarterly,41*(1), 65–73. https://doi.org/10.1177/107769906404100108

Tantillo, J., Di Lorenzo-Aiss, J., & Mathisen, R. E. (1995). Quantifying perceived differences in type styles: An exploratory study. *Psychology and Marketing,12*(5), 447–457. https://doi.org/10.1002/mar.4220120508

Tapscott, D. (1998). *Growing up digital: The rise of the Net Generation*. McGraw-Hill.

Tarita-Nistor, L., Lam, D., Brent, M. H., Steinbach, M. J., & González, E. G. (2013). Courier: A better font for reading with age-related macular degeneration. *Canadian Journal of Ophthalmology,48*(1), 56–62. https://doi.org/10.1016/j.jcjo.2012.09.017

Taylor, C. D. (1934). The relative legibility of black and white print. *Journal of Educational Psychology,25*(8), 561–578. https://doi.org/10.1037/h0074746

Taylor, J. L. (1990). *The effect of typeface on reading rates and the typeface preferences of individual readers* (Publication No. 303890933) [Doctoral dissertation, Wayne State University]. ProQuest Dissertations and Theses Global.

Thompson, G. B. (2009). The long learning route to abstract letter units. *Cognitive Neuropsychology,26*(1), 50–69. https://doi.org/10.1080/02643290802200838

Tinker, M. A. (1944). Criteria for determining the readability of type faces. *Journal of Educational Psychology,35*(7), 385–396. https://doi.org/10.1037/h0055211

Tinker, M. A. (1963). *Legibility of print*. Iowa State University Press.

Tinker, M. A. (1966). Experimental studies on the legibility of print: An annotated bibliography. *Reading Research Quarterly, 1*(4), 67–118. https://www.jstor.org/stable/747222

Tinker, M. A., & Paterson, D. G. (1942). Reader preferences and typography. *Journal of Applied Psychology,26*(1), 38–40. https://doi.org/10.1037/h0061105

Treiman, R., Gordon, J., Boada, R., Peterson, R. L., & Pennington, B. F. (2014). Statistical learning, letter reversals, and reading. *Scientific Studies of Reading,18*(6), 383–394. https://doi.org/10.1080/10888438.2013.873937

Tullis, T. S., Boynton, J. L., & Hersh, H. (1995). Readability of fonts in the Windows environment. In I. Katz, R. Mack, & L. Marks (Ed.), *CHI'95 Conference Companion on Human Factors in Computing Systems* (pp. 127–128). Association for Computing Machinery. https://doi.org/10.1145/223355.223463

Tyrrell, R. A., Pasquale, T. B., Aten, T., & Francis, E. L. (2001). Empirical evaluation of user responses to reading text rendered using ClearType^{TM} technologies. *Society for Information Display International Symposium Digest of Technical Papers,32*(1), 1205–1207. https://doi.org/10.1889/1.1831776

Uman, L. S. (2011). Systematic reviews and meta-analyses. *Journal of the Canadian Academy of Child and Adolescent Psychiatry,20*(1), 57–59. https://doi.org/10.1002/9781444311723.ch8

Universal design: Ensuring access to the general education curriculum. (1999, Fall). *Research Connections in Special Education*, No. 5, pp. 1–5. ERIC. https://files.eric.ed.gov/fulltext/ED433666.pdf

Uysal, S. A., & Düger, T. (2012). Writing and reading training effects on font type and size preferences by students with low vision. *Perceptual and Motor Skills, 114*(3), 837–846. https://doi.org/10.2466/15.10.11.24.PMS.114.3.837-846

Vaidya, C. J., Gabrieli, J. D. E., Verfaellie, M., Fleischman, D., & Askari, N. (1998). Font-specific priming following global amnesia and occipital lobe damage. *Neuropsychology, 12*(2), 183–192. https://doi.org/10.1037/0894-4105.12.2.183

Vanderplas, J. M., & Vanderplas, J. H. (1980). Some factors affecting legibility of printed materials for older adults. *Perceptual and Motor Skills, 50*(3, Pt. 1), 923–932. https://doi.org/10.2466/pms.1980.50.3.923

Vernon, M. D. (1929). The errors made in reading. In *Studies in the psychology of reading* (pp. 5–40) (Reports of the Committee upon the Physiology of Vision, III. Privy Council, Medical Research Council, Special Report Series No. 130). His Majesty's Stationery Office.

Vernon, M. D. (1931). *The experimental study of reading*. Cambridge University Press.

Wade, N. J., & Tatler, B. W. (2008). Did Javal measure eye movements during reading? *Journal of Eye Movement Research, 2*(5), 1–7. https://doi.org/10.16910/jemr.2.5.5

Walker, S. (2013). *Book design for children's reading: Typography, pictures, print*. St Bride Foundation.

Walker, S., & Reynolds, L. (2003). Serifs, sans serifs and infant characters in children's reading books. *Information Design Journal, 11*(2–3), 106–122. https://doi.org/10.1075/idj.11.2.04wal

Wang, P., & Nikolić, D. (2011). An LCD monitor with sufficiently precise timing for research in vision. *Frontiers in Human Neuroscience, 5*, Article 85. https://doi.org/10.3389/fnhum.2011.00085

Watanabe, R. K., Gilbreath, M. K., & Sakamoto, G. C. (1994). The ability of the geriatric population to read labels on over-the-counter medication containers. *Journal of the American Optometric Association, 65*(1), 32–37.

Watts, L., & Nisbet, J. (1974). *Legibility in children's books: A review of research*. NFER.

Weaver, D. F. (2014). Font specific reading-induced seizures. *Clinical Neurology and Neurosurgery, 125*, 210–211. https://doi.org/10.1016/j.clineuro.2014.07.035

Weaver, D. F., & Hawco, C. L. A. (2015). Geminate consonant grapheme-colour synaesthesia (ideaesthesia). *BMC Neurology, 15*, ArtID 112. https://doi.org/10.1186/s12883-015-0372-7

Webster, H. A., & Tinker, M. A. (1935). The influence of type face on the legibility of print. *Journal of Applied Psychology, 19*(1), 43–52. https://doi.org/10.1037/h0063604

Weiss, A. P. (1917). The focal variator. *Journal of Experimental Psychology, 2*(2), 106–113. https://doi.org/10.1037/h0072089

Weiss, M. J. (1978). *Children's preferences in tradebook format factors* (ED188139). ERIC. https://files.eric.ed.gov/fulltext/ED188139.pdf

Weiss, M. J. (1982). Children's preferences for format factors in books. *The Reading Teacher, 35*(4), 400–406. https://www.jstor.org/stable/20198002

Wendt, D. (1968). Semantic differentials of typefaces as a method of congeniality research. *Journal of Typographic Research, 2*(1), 3–25.

Wendt, D. (1969). *Einflüsse von Schriftart (Bodoni vs. Futura), Schriftneigung und Fettigkeit auf die erzielbare Lesegeschwindigkeit mit einer Druckschrift* (Untersuchungen zur Lesbarkeit von Druckschriften, Bericht 5) [Influences of typeface (Bodoni vs. Futura), font, and stroke thickness on achievable reading speed with a leaflet (Studies on the Legibility of Printed Material, Report No. 5]. Universität Hamburg, Psychologisches Institut.

Wendt, D. (1994). Legibility. In P. Karow, *Font technology: Methods and tools* (pp. 271–306). Springer. https://doi.org/10.1007/978-3-642-78505-4_12

Wheildon, C. (1984). *Communicating, or just making pretty shapes? A study of the validity—or otherwise—of some elements of typographic design.* Newspaper Advertising Bureau of Australia.

Wheildon, C. (1990). *Communicating or just making pretty shapes?* (3rd ed.). Newspaper Advertising Bureau of Australia.

Wheildon, C. (with Heard, G., & Ogilvy, D.) (2005). *Type & layout: Are you communicating or just making pretty shapes.* Worsley Press.

Wheildon, C. (with Ogilvy, D., Warwick, M., & Antin, T.) (1995). *Type & layout: How typography and design can get your message across—or get in the way.* Strathmore Press.

Whittemore, I. C. (1948). What do you mean, legibility? *Print,5*(4), 35–37.

Wiebelt, A. (2004). Do symmetrical letter pairs affect readability? A cross-linguistic examination of writing systems with specific reference to the runes. *Written Language and Literacy,7*(2), 275–304. https://doi.org/10.1075/wll.7.2.07wie

Wilkins, A. (1986). Intermittent illumination from visual display units and fluorescent lighting affects movements of the eyes across text. *Human Factors,28*(1), 75–81. https://doi.org/10.1177/001872088602800108

Wilkins, A., Cleave, R., Grayson, N., & Wilson, L. (2009). Typography for children may be inappropriately designed. *Journal of Research in Reading,32*(4), 402–412. https://doi.org/10.1111/j.1467-9817.2009.01402.x

Wilkins, A., Nimmo-Smith, I., Tait, A., McManus, C., Della Sala, S., Tilley, A., Arnold, K., Barrie, M., & Scott, S. (1984). A neurological basis for visual discomfort. *Brain,107*(4), 989–1017. https://doi.org/10.1093/brain/107.4.989

Wilkins, A., Smith, K., & Penaccio, O. (2020). The influence of typography on algorithms that predict the speed and comfort of reading. *Vision, 4*(1), Article No. 18. https://doi.org/10.3390/vision4010018

Wilkins, A. J. (2021). Visual stress: Origins and treatment. *CNS, 6*(1), 74–86. https://www.oruen.com/journals/january-2021/visual-stress-origins-and-treatment

Wilkins, A. J., Jeanes, R. J., Pumfrey, P. D., & Laskier, M. (1996). Rate of Reading Test®: Its reliability, and its validity in the assessment of the effects of coloured overlays. *Ophthalmic and Physiological Optics,16*(6), 491–497. https://doi.org/10.1016/0275-5408(96)00028-2

Wilkins, A. J., & Nimmo-Smith, M. I. (1987). The clarity and comfort of printed text. *Ergonomics,30*(12), 1705–1720. https://doi.org/10.1080/00140138708966059

Wilkins, A. J., Smith, J., Willison, C. K., Beare, T., Boyd, A., Hardy, G., Mell, L., Peach, C., & Harper, S. (2007). Stripes within words affect reading. *Perception,36*(12), 1788–1803. https://doi.org/10.1068/p5651

Williams, M. A. (1990). *Legibility of serif and sans serif type faces in computer displays* [Master's thesis, Colorado State University]. Retrieved June 30, 2021, from https://mountainscholar.org/handle/10217/199725.

Williamson, H. (1966). *Methods of book design: The practice of an industrial craft* (2nd ed.). Oxford University Press.

Wilson, L., & Read, J. (2016). Do particular design features assist people with aphasia to comprehend text? An exploratory study. *International Journal of Language and Communication Disorders,51*(3), 346–354. https://doi.org/10.1111/1460-6984.12206

Woods, R. J., Davis, K., & Scharff, L. F. V. (2005). Effects of typeface and font size on legibility for children. *American Journal of Psychological Research, 1*(1), 86–102. Retrieved June 30, 2021, from https://www.mcneese.edu/wp-content/uploads/2020/08/ajpr9.pdf.

Xiong, Y.-Z., Lorsung, E. A., Mansfield, J. S., Bigelow, C., & Legge, G. E. (2018). Fonts designed for macular degeneration: Impact on reading. *Investigative Ophthalmology and Visual Science,59*(10), 4182–4189. https://doi.org/10.1167/iovs.18-24334

Yager, D., Aquilante, K., & Plass, R. (1998). High and low luminance letters, acuity reserve, and font effects on reading speed. *Vision Research,38*(17), 2527–2531. https://doi.org/10.1016/S0042-6989(98)00116-3

Yule, V. (1988). The design of print for children: Sales-appeal and user-appeal. *Reading,22*(2), 96–105. https://doi.org/10.1111/j.1467-9345.1988.tb00662.x

Zachrisson, B. (1965). *Studies in the legibility of printed text*. Almqvist & Wiksell.

Zuccollo, G., & Liddell, H. (1985). The elderly and the medication label: Doing it better. *Age and Ageing,14*(6), 371–376. https://doi.org/10.1093/ageing/14.6.371

Author Index

A
Adams, M. J., 24
Adams, S., 86
Akhmadeeva, L., 37, 38, 84
Alexander, P. A., 84
Ali, A. Z. M., 119
Anderson, J. L., 71
Aquilante, K., 109
Arditi, A., 24, 36, 78, 84, 92, 103, 110, 111
Arnold, E. C., 44
Aten, T. R., 89

B
Baethge, C., 79
Baeza-Yates, R., 109, 110
Bailey, I. L., 32
Balfour, S., 59
Banerjee, J., 98
Baron, N. S., 84
Bartram, D., 40, 49
Bhattacharyya, M., 99
Beier, S., 31, 36, 84, 95
Bell, R. C., 35
Bennett, A. G., 32
Benton, C. L., 39
Berliner, A., 39
Bernard, J.-B., 111
Bernard, M. L., 103, 114–117
Beymer, D., 119
Beyon, J., 120
Bigelow, C. A., 4, 15, 84, 85, 103
Bjørndal, A., 8
Bluhm, A., 54
Boyarski, D., 100, 104
Branch, R. M., 114
Braun, C. C., 60, 61, 70

Bringhurst, R., 2
British Dyslexia Association, 71, 73, 109, 110
British Standards Institute, 32
Brockington, G., 53
Brookshire, C. E., 71
Brown, C. H., 37
Brown, E., 89
Brumberger, E. R., 49
Burt, C., 24, 53–55, 61, 75

C
Caldera, C. I., 98, 104
Campbell, K. A., 69, 71
Caplan, P. J., 9
Catich, E. M., 3
Cattell, J. M., 13
Chaparro, B. S., 83, 92, 94, 103, 116, 117, 119
Chernecky, C., 118
Child, I. L., 87
Chomsky, N., 11
Chung, S. T. L., 9
Click, J. W., 45
Clough, J., 5, 6
Coghill, V., 53, 57, 58
Cossu, G., 59
Cowan, A., 32
Cox-Boyd, C., 120
Craig, J., 17, 21
Crist, W. B., 59
Csilla, K. P., 101

D
Dale, N., 53, 54

Davidow, A., 113
Davidson, H. F., 59
Davis, R. C., 39
De Bruijn, O., 103
Deederer, C., 32
De Lange, R. W., 57
Dillon, A., 89
Dreyfus, J., 22, 38
Düger, T., 68, 76
Dyson, M. C., 31, 115

E
Earnest, W. J., 87
English, E., 44
Errando Oyonarte, C. L., 22, 23
Estey, A., 68, 71

F
Flynne, L. P., 45
Fox, D., 94, 118
Fox, J. G., 37
Franken, R. B., 39
Frank, T., 117
Franz, S. I., 27

G
Gallagher, T. J., 54
Garcia, M. L., 98
Garon, J. E., 87
Gasser, M., 36, 50, 120
Gilbert, L. C., 102
Goikoetxea, E., 59
González-Rodriguez, R., 22, 23
Gosse, R., 113
Gould, J. D., 88, 97
Grant, M. M., 114
Gray, N., 6
Griffing, H., 27
Griffiths, G. W., 58
Grooters, L. E., 86
Gross, J. S., 50
Gugerty, L., 89

H
Halcomb, C. G., 115, 117
Hamai, M., 21
Hargreaves, A., 58
Harris, J., 31, 95
Hartley, J., 37, 55
Haskins, J. B., 44, 45

Haugen, T. T., 67
Haw, C., 71
Hawco, C. L. A., 24
Hawking, S. W., 101
Hearnshaw, L. S., 55
Hedges, L. V., 9
Hedlich, C., 70
Herbert, R., 71
Hering, E., 4
Hetherington, R., 32
Hofstätter, P. R., 38, 39
Hojjati, N., 101, 104, 124
Holleran, P. A., 104
Holmes, K., 84, 103
Howe, N., 83
Hutner, N., 91
Hvistendahl, J. K., 43
Hwang, W. S., 91

I
Ing, E., 87

J
Jacklin, C. N., 9
Jacobson, W. S., 54
Jainta, S., 93
Jelfs, A., 84
Jorgensen, B., 83
Josephson, S., 98, 99
Juni, S., 50

K
Kahl, M. R., 43, 44
Kaspar, K., 106, 125
Kastl, A. J., 87
Kempson, E., 46
Kennedy, H., 59
Kerr, J., 54, 55, 61
Kim, M., 28, 104
Kinross, R., 6, 12, 13, 55
Koch, B. E., 105
Komoda, M. K., 91, 94, 102
Kong, Y.-K., 28, 92
Kosztyán, Z. T., 120
Krulee, G. K., 24
Kullmann, D. M., 23, 77
Kuster, S. M., 72, 73

L
Lamare, M., 4

Larson, K., 36, 84, 95
Legge, G. E., 15
Legros, L. A., 31, 59, 95, 123
Lenze, J. S., 100, 104
Liao, C. H., 116
Lida, B., 115
Liddell, H., 60
Lightfoot, C., 5
Ling, J., 118
Liversedge, S. P., 93
Lockhead, G. R., 59
Lovie, J. E., 32
Luckiesh, M., 12, 28
Lund, O., 6, 28

M
Maccoby, E. E., 9
Mach, E., 59
Mackiewicz, J., 49
Marinus, E., 72
Mátrai, R., 120
Mauretto, P., 104
McCarthy, M. S., 49, 88
McKown, J., 117
McLean, R., 6, 21, 59
McVey, G. F., 88
Mertens, S., 79
Mills, M. M., 115–117
Misanchuk, E. R., 104
Mizrachi, D., 84
Moeller, R., 49
Mohammadi, E., 37
Mollen, J. D., 91
Moore, N., 46
Moret-Tatay, C., 84, 93
Moriarty, S. E., 44
Morison, S., 4, 21, 22, 77
Morrison, G. R., 40
Morris, R. A., 103
Mosley, J., 2, 4–6
Moss, F. K., 12, 28
Mothersbaugh, D. L., 49
Muniandy, B., 101, 104, 124
Muter, P., 104
Myung, R., 117

N
Nersveen, J., 70, 71
Nikolić, D., 91
Nimmo-Smith, M. I., 130
Nisbet, J., 54

Nolan, C. Y., 63, 75
Noonan, E., 8
Novy, F., 24
Nutt, D. J., 22

O
Olkin, I., 9
Osgood, C. E., 38–40
Ovink, G. W., 6, 28, 35, 39

P
Park, K.-S., 104
Paterson, D. G., 14, 27, 38, 57, 78
Pedró, F., 83
Penrose, R., 101
Perea, M., 6, 77, 84, 93, 99, 103, 110
Peterson, M., 115
Phillips, R. M., 85–87
Pittman, B., 66, 75
Plass, B., 109
Poffenberger, A. T., 39
Polden, P. G., 91
Poncelet, G. M., 85, 126
Poole, A., 46, 47
Popper, K. R., 130
Popp, H. M., 59
Potter, M. C., 59
Poulton, E. C., 16, 17, 37
Powell, S. L., 72, 73
Prensky, M., 83
Prince, J. H., 68
Proctor, L. F., 85, 126
Pyke, R. L., 4, 12, 14, 15, 27, 47

R
Raban, B., 54
Raymaker, D. M., 126
Rayner, K., 89
Read, J., 71, 75
Redich, J., 113
Rello, L., 109, 110
Reynolds, L., 12, 36, 53, 54, 58
Richardson, J. T. E., 84
Richards, O. W., 32, 84
Ripoli, J. C., 58, 59
Robinson, D. O., 4
Roethlein, B. E., 27
Rooum, D., 55
Rose, T. A., 71
Rowe, C. L., 39, 49
Rubin, G. S., 69, 75, 102

Russell, M., 103
Russell-Minda, E., 78, 79

S

Sanocki, T., 24, 84
Sassoon, R., 57, 66
Savory, P., 104
Scheiner, E. C., 44
Schiffman, H. R., 4
Schiller, G., 39
Schriver, K. A., 43, 46, 48, 54, 76, 85, 88, 126
Schwarz, N., 15
Schweinberger, S. R., 24
Seale, J., 1
Shaikh, A. D., 83, 105, 118, 119
Shaw, A., 64, 68
Shaw, S. C. K., 71
Sheedy, J. E., 88, 89
Singer, L. M., 84
Skilton, A., 68
Slattery, T. J., 89
Sloan, L. L., 32
Smither, J. A.-A., 60, 61, 70
Smith, H. J., 39
Snellen, H., 32
Soleimani, H., 37
Song, H., 15
Stempel, G. H., III., 45
Stiff, P., 6
Stone, D. B., 98
Storrer, K., 115
Strauss, W., 83
Suen, C. Y., 91, 94, 102
Sullivan, J. L. F., 35
Svensson, E., 29

T

Tannenbaum, P. H., 39, 40
Tantillo, J., 40
Tapscott, D., 83
Tarita-Nistor, L., 69
Tatler, B. W., 4
Taylor, C. D., 4
Taylor, J. L., 35
Thompson, G. B., 59
Tinker, M. A., 12–14, 27, 28, 31, 38, 44, 57, 78, 95
Treiman, R., 60
Trice, A. D., 72, 73

Tullis, T. S., 97, 104
Turano, K., 102
Tyrrell, R. A., 89

U

Uman, L. S., 8
Uysal, S. A., 68, 76

V

Vaidya, C. J., 24
Vanderplas, J. H., 60
Vanderplas, J. M., 60
Van Schaik, P., 118
Vernon, M. D., 31, 95, 123

W

Wade, N. J., 4
Walker, S., 53, 54, 58
Wang, P., 91
Watanabe, R. K., 60
Watts, L., 53
Weaver, D. F., 24
Webster, H. A., 28
Weiss, A. P., 12, 56
Weiss, M. J., 56
Wendt, D., 35, 39
Wheildon, C., 45–48, 51, 75
Whittemore, I. C., 43
Wiebelt, A., 59
Wilkins, A. J., 29–31, 36, 57, 89, 93, 94, 129, 130
Williams, M. A., 100
Williamson, H., 17, 21
Wilson, L., 71, 75
Woods, R. J., 24, 86

X

Xiong, Y.-Z., 111

Y

Yager, D., 102, 109
Yule, V., 59

Z

Zachrisson, B., 12, 14, 43, 55, 56, 61, 76
Zuccollo, G., 60

Subject Index

A

Advertising products, 39
Age-related macular degeneration, *see* macular degeneration
Aliasing, 88, 89, 92
Amazon Kindles, 101
American Psychological Association, *Publication Manual*, 77, 79
Anti-aliasing software, 88, 89
Antique typefaces, 6, 63
APA PsycInfo (database), 8
Aphasia, 71, 73, 75
Artificial typefaces, 31, 37, 49, 84, 85, 95, 123, 125
Ascenders, 16, 30, 59, 85
Atmosphere value, 38
Autocorrelation, horizontal, 29, 30, 33, 93–95, 129

B

Backward masking, 92, 94
Backward searching, 9
Bibliographic searches, 8
Binocular rivalry, 14
Block typefaces, 27. *See also* Slab serif typefaces
Body size, 15–17, 70, 85, 100
Box scores, 9
Brain (journal), 23

C

Cap-height, 16
Carolingian minuscule, 3, 4
Carry-over effects, *see* Transfer effects
Cataract, 65, 68–70

Cathode-Ray Tubes (CRTs), 13, 88–93, 97, 100, 102, 103, 110, 111
Cellular phones, 67, 103
Chicago Manual of Style, 2
Children
  reading from paper, 53–60
  reading from screens, 84, 85, 117
ClearType software, 89, 92, 105
Comprehending text
  reading from paper, 36
  reading from screens, 99
Comprehension versus factual recall, 37
Comprehension versus reading, 101
Computational models, 4
Confusions among letters
  reading from paper, 31
  reading from screens, 94
Conjectures and refutations, 130
Connectionist models, 4
Connotations of typefaces
  reading from paper, 39
  reading from screens, 104
Connotative meaning, 38, 39, 41
Context, 1, 9, 32, 43, 47, 48, 51, 76, 78, 79, 95, 114, 116, 117, 125
Cursive typefaces, 8, 15, 49, 58, 59, 66, 71, 98, 105, 121, 124
Cyrillic script, 37

D

Deaf-blindness, 68
Descenders, 16, 30, 67, 85
Designers' attitudes, 85, 126
Digital Natives, 83, 90
Direction signs, 6
Display typefaces, 7, 105

Subject Index

Distance method, 12, 14, 27, 28, 31
Dyslexia
   reading from paper, 71
   reading from screens, 109

E
Early typography, 6
Ecological validity, 13, 103
Egyptian typefaces, 32, 40
Epileptic seizures, 24
ERIC (database), 8
"Everybody knows", 79, 126
Expectations, 43, 51, 76, 107, 125, 126
Eye movements
   reading from paper, 4, 13, 23, 35, 65
   reading from screens, 89, 99, 102, 104, 106, 109, 110, 112, 119, 124

F
Familiarity (with typefaces), 13, 107, 125
Fatigue in reading, 13. *See also* transfer effects
Fonts versus typefaces, 17
Forward searching, 9
Fourier analysis, 130

G
Generation Y, 83
Glaucoma, 65, 69
Gothic (gothic) typefaces, 27, 28, 32, 37, 40, 60, 66, 91, 92, 97, 102
Grapheme–colour synaesthesia, 24
Gravity chronometer, 13
Grotesque typefaces, 6

H
Handheld devices, 103, 104, 107
Headlines, 44–48, 51

I
Infant characters, 53, 58
Internet browsers, 113, 117, 118, 121, 124, 126
Irradiation, 4, 55, 129

J
Journal of Psychopharmacology, 22

K
Korean, 28, 75, 91, 92, 104, 117, 124

L
Leading, 16, 46, 93
Legibility, 2, 4, 8–17, 21–23, 27–29, 32, 33, 35, 36, 38, 40, 41, 43, 45, 47–49, 51, 55–59, 61, 66, 68, 69, 71, 73, 75, 77–79, 84–87, 89, 90, 92, 94, 95, 97, 100–102, 104–106, 110, 112–116, 119–121, 123, 124, 126, 127, 129, 130
Legibility, methods for measuring, 12–14
Legibility versus readability, 11, 17
Letter reversals, 59
Lexical decisión, 93
Liquid Crystal Displays (LCDs), 13, 88–95, 98, 101, 110, 111, 113, 114, 119, 130

M
Macular degeneration, 65, 69, 110, 124
Medication labels, 61
MEDLINE (database), 8
Merriam-Webster's Manual for Writers and Editors, 22, 77
Meta-analysis, 9, 75, 123
Michelson contrast, 130
Millenials, 83
Ming typefaces, 28, 92
Mobile phones, 103
Myopia, 65

N
Narrative reviews, 8, 9
Net Generation, 83

O
Older adults
   reading from paper, 59–61
   reading from screens, 116
Overhead projectors, 29, 85–87, 90

P
Palmtop computers, 103
Palo seco [dry stick], 22, 23
Peripheral vision, 12, 56
Personal digital assistants, 103, 104, 124
Personality, 40, 116, 119, 125
Personality, of typefaces, 38, 49, 107

Polarity profile, 38
PowerPoint, 12, 87, 90
Practice effects, *see* Transfer effects
Preferences
    reading from paper, 76
    reading from screens, 125
Printing screen-based text to hard copy, 2

Q
Quotation errors, 79

R
Rapid serial visual presentation, 102, 103, 107, 109, 112, 124, 126, 127
Reading letters and words
    reading from paper, 27
    reading from screens, 91
Reading text
    reading from paper, 35
    reading from screens, 97
Reflex blink method, 12
Researcher bias, 48, 67
Revista Española de Anestesiología y Reanimación, 22
Roman Empire, 2–4
Roman Republic, 5
Roman (roman) typefaces, 4

S
Sans serif inscriptions, 6
Sans serif typefaces, defined, 4
Scale-to-grey, 88, 89
Semantic differential, 38, 39, 41, 45
Serif inscriptions, 5, 6
Serif typefaces, defined, 2

Short-exposure method, 12, 13, 27, 28, 35
Size of typefaces, 15, 17, 85
Slab serif typefaces, defined, 6
Slide projectors, 85–87, 90
Smartphones, 28, 103, 104, 107, 113, 124
Snellen chart, 6, 32
Spanish-speaking countries, 58
Special education, 66, 67
Speed–accuracy trade-off, 116
Speed of reading, 13–15, 117
Subjective reports, 14, 17
Systematic reviews, 8–10

T
Tachistoscopic presentation, 123
Transfer effects, 120
Typeface-specific information, 24
Type I errors, 50, 66, 106
Typographers' attitudes, 21

U
Universal design, 85, 126

V
Visibility thresholds, 12, 14, 28
Visual acuity, 6, 32, 33, 64, 103
Visual impairment, acquired, 68, 73
Visual impairment, congenital, 68, 73, 75
Vote counting, 9

X
X-height, 16, 17, 30, 38, 48, 49, 57, 58, 64, 70, 72, 85, 100, 102

Typeface Index

A
Adsans, 69, 79
Agency FB, 115
Akzidenz-Grotesk, 23
American Typewriter, 78
Andale Mono, 69
Antique with Old Style, 63
Arial, 3, 5, 15, 24, 30, 37, 50, 58, 59, 68–70, 72, 77–79, 84, 87, 93, 97–99, 101, 106, 109, 110, 114–120, 126
Avant Garde, 113
Avant Garde Gothic, 40

B
Baskerville, 2, 3, 31, 37
Batang, 104, 117
Bauer Bodoni, 48
Bauhaus Md BT, 49
Bembo, 37, 55
Block, 6, 22, 27, 29, 32, 87, 93, 99
Bodoni, 28, 35, 39, 44, 48, 60
Book Antiqua, 101
Bookman, 67
Bookman Old Style, 37
Bradley Hand ITC, 115
Brush455 BT, 15
Bulletin, 86

C
Calibri, 77, 78, 92, 101
Cambria, 92
Candara, 92
Centaur, 94
Century, 2–6, 11–13, 21, 22, 25, 32, 33, 44, 58, 59, 85, 86, 103

Century Schoolbook, 40, 60, 61, 68, 113, 115, 116
Clearview, 69
Clearview Text, 94
Cloister Black, 78
Comic Sans, 3, 58, 59, 67, 68, 115–117
Computer Modern, 77
Computer Modern Unicode, 110
Consolas, 92, 123
Constantia, 92, 94
Corbel, 92
Corona, 46
Coronet, 45
CounselorScript, 49
Courier, 50, 60, 61, 69, 70, 91, 98, 102, 111, 113, 115, 116, 123
Courier New, 70, 99, 113, 115–117, 120

D
Dodum, 28, 29, 104, 117, 118
Droid, 103
Dutch, 102, 103
Dyslexie, 72, 73

E
Egyptian, 6, 40
Egyptian Paragon, 32
Eido, 111
Elite, 37, 86

F
Fin Grotesk, 55
Folio, 39
Foundry Form Sans, 69

Franklin Gothic, 27
Frutiger, 70
Futura, 35, 39, 40, 43–45, 48, 56, 57

G
Garamond, 2, 3, 39, 45, 48, 94, 101, 110
Geneva, 30
Georgia, 77, 84, 98, 100, 101, 115, 116, 119, 120
Gill Sans, 35, 55, 57, 58, 64, 66
Glypha, 105
Gothic, 6, 27, 28, 32, 40, 60, 66, 86, 91, 92, 97, 102, 123
Gothic Elite, 37
Goudy Old Style, 40, 115
Grotesk, 14, 55
Grotesque 215, 37
Gulim, 28, 104, 117, 118
Gungseo, 28, 29, 104

H
Hallmarke, 113
Harrington, 71
Helvetica, 35, 39, 40, 43, 44, 46, 48, 50, 57, 58, 60, 61, 67, 69, 70, 79, 98, 100, 102, 105, 110, 111, 113, 114, 119

I
Imperial, 6, 43
Imprint, 14

K
Kabel, 27, 28, 39, 78
Karnak, 44

L
LeRoy Standard, 86
LeRoy Stymie, 86
Letter Gothic, 91, 97, 102, 123
Lexia Readable, 58, 59
Lining Grotesque, 14, 27
Lo, 28
Lucida, 69, 99, 103
Lucida Bright, 85, 93, 106
Lucida Casual, 98
Lucida Console, 120
Lucida Sans, 30, 77, 93, 99, 106

M
Mager Futura, 56
Mager Konsul, 55
Maxular Rx, 111, 112
Mediaeval, 56
Metrolite Medium, 63
Metrolite No. 2, 28
Mistral, 15
Modern, 37, 40
Monaco, 50
Monotype Corsiva, 115
MS Sans Serif, 97, 98
MS Serif, 97
Myriad, 110

N
New Century Schoolbook, 113
News #2, 43
News Gothic, 27
News Sans, 43
Nordisk Antikva, 55

O
Old Style, 37, 40, 63, 115
Optima, 48

P
Paladium, 57
Palatino, 2, 3, 24, 30, 39, 48, 50, 93, 113, 120
Palatino Linotype, 120
Parinesy, 57
Perpetua, 16, 17
Pica, 66
Plantin, 64, 66
Poster Bodoni, 28
Press, 11, 35, 97
Press Roman, 35
Primary, 30, 33, 40, 51, 57, 64, 66, 68

Q
Quick Type, 113

R
Roboto, 103
Rockwell, 94
Royal, 43, 69

S

San Francisco, 103
Sans Heavy, 43
Sassoon Primary, 30, 57
Scala, 70
Scala Sans, 70
Scotch Roman, 78
Script, 3, 37, 45, 66, 97
Small Font, 97
Spartan, 39, 45, 60
Spartan Black, 45
Standard Elite, 37
Swiss, 102
Sylfaen, 58, 59
System, 1, 4, 6, 12, 67, 72, 73, 83, 87, 90, 97, 98, 102, 110–112, 114, 124, 129

T

Tahoma, 3, 5, 30, 68, 99, 115, 116
Tempo, 44

Times New Roman, 2, 3, 4, 16, 17, 21, 24, 29, 30, 35, 40, 49, 50, 58, 59, 68, 69, 72, 77, 78, 79, 84, 87, 92, 98, 99, 101, 106, 114–120, 123
Times Roman (Times), 4, 35, 39, 44, 48, 57, 58, 60, 67, 69, 70, 84, 98, 100, 102, 109, 111
Times Sans Serif, 29
Tiresias, 69, 70
Trade Gothic, 60
Transitional, 40
Twentieth Century, 13, 22, 44, 86, 103

U

Univers, 16, 17, 31, 35, 37, 40, 48, 68, 97

V

Verdana, 3, 5, 24, 30, 57, 68–71, 79, 84, 87, 89, 92–94, 98–101, 110, 115, 116, 118–120, 123

The manufacturer's authorised representative in the EU is Springer Nature Customer Service Centre GmbH, Europaplatz 3, 69115 Heidelberg, Germany. If you have any concerns regarding our products, please contact ProductSafety@springernature.com

Printed and bound by CPI Group (UK) Ltd, Croydon, CR0 4YY

25/03/2026

02078197-0014